D0287551

Human Nutrition

Human Nutrition

A continuing debate

An Account of a Symposium
Nutrition in the Nineties
which formed part of the Edinburgh Science Festival

The Symposium was organized by the Pfizer Symposium
Committee of the University of Edinburgh
Post-Graduate Board for Medicine

MARTIN EASTWOOD
Gastrointestinal Unit, Department of Medicine, University of Edinburgh, Western General Hospital, Edinburgh

CHRISTINE EDWARDS
Department of Human Nutrition, University of Glasgow

DOREEN PARRY
Department of Nutrition, Queen Margaret College, Edinburgh

Published by Chapman & Hall, 2–6 Boundary Row, London SE1 8HN

Chapman & Hall, 2–6 Boundary Row, London SE1 8HN, UK

Van Nostrand Reinhold Inc., 115 5th Avenue, New York NY10003, USA

Chapman & Hall Japan, Thomson Publishing Japan, Hirakawacho Nemoto Building, 7F, 1-7-11 Hirakawa-cho, Chiyoda-ku, Tokyo 102, Japan

Chapman & Hall Australia, Thomas Nelson Australia, 102 Dodds Street, South Melbourne, Victoria 3205, Australia

Chapman & Hall India, R. Seshadri, 32 Second Main Road, CIT East, Madras 600 035, India

First edition 1992

Typeset in 10/12 pt Ehrhardt by Best-set Typesetters Ltd, Hong Kong
Printed in Great Britain by TJ Press (Padstow) Ltd, Padstow, Cornwall

ISBN 0 412 40310 2 0 442 31445 0 (USA)

A catalogue record for this book is available from the British Library

Library of Congress Cataloging-in-Publication data available

Contents

Contributors

Mr K.G. Anderson, Convenor of Good Manufacturing Practices Panel at the Institute of Food Science and Technology, and joint editor of the *IFST Guide to the Responsible Management of GMP*

Professor D.J.P. Barker, Director of the MRC Environmental Epidemiology Unit, University of Southampton

Dr D.G. Beevers, Reader in Medicine, University of Birmingham, Dudley Road Hospital, Birmingham

Professor A.E. Bender, formerly Professor of Nutrition, University of London

Dr S.A. Bingham, MRC Dunn Nutrition Centre, University of Cambridge

Dr J.H. Cummings, MRC Dunn Nutrition Centre, University of Cambridge

Dr M.A. Eastwood, Reader in Medicine, Western General Hospital, University of Edinburgh

Dr Christine A. Edwards, Research Fellow, Department of Medicine, Western General Hospital, University of Edinburgh

Professor J.S. Garrow, Professor of Nutrition, University of London

Professor J.G.A.J. Hautvast, Professor of Nutrition, University of Wageningen, The Netherlands

Dr K.W. Heaton, Reader in Medicine, Bristol Royal Infirmary, University of Bristol

Professor A.A. Jackson, Professor of Human Nutrition, University of Southampton

Professor W.P.T. James, Director of the Rowett Research Institute, Aberdeen

Dr D. Kritchevsky, Associate Director, The Wistar Institute, Philadelphia, USA

Dr D.A. McCarron, Oregon Health Sciences University, Oregon, USA

Professor I. Macdonald, Professor of Physiology, Guy's Hospital, University of London

Professor Doreen Parry, Queen Margaret College, Edinburgh

Dr C.J. Packard, Royal Infirmary, Glasgow

Dr R. Passmore, Senior editor of *Davidson and Passmore Textbook of Nutrition*

Ann Ralph, Rowett Research Institute, Aberdeen

Molly E. Reusser, Oregon Health Sciences University, Oregon, USA

Professor I.H. Rosenberg, Director, USDA Human Nutrition Research Center on Ageing at Tufts University, Boston, USA

Professor J. Shepherd, Professor of Clinical Biochemistry, Royal Infirmary, Glasgow

Professor B.A. Wharton, Professor of Nutrition and Paediatrics, University of Glasgow

Foreword

There is, as in Philosophy, so in Divinity, sturdy doubts and boisterous objections, wherewith the unhappiness of our knowledge too nearly acquainted is.
Sir Thomas Browne *Religio Medici*

This was written in 1635 by a young doctor, recently returned from his studies in Padua, Montpelier and Leiden, who was about to start a long career as a physician in Norwich. The 'sturdy doubts' and 'boisterous objections' of the time were soon to lead to a civil war and the execution of the king, Charles I. A young doctor today because of the 'unhappiness of our knowledge' of the science of nutrition should have similar doubts about what advice and how much advice he should give to his patients about the choice of their foods. Perhaps the advice that he gives may spare some of his patients from premature death.

This book records the papers presented at a conference in Edinburgh, organized by bodies responsible for postgraduate medicine, and subsequent discussions in which their audience took part. There were sixteen speakers, twelve from the UK, three from the USA and one from the Netherlands, each of whom has had a long experience of nutritional problems.

Nutrition in the 1990s will be concerned with problems arising from the progressive changes over the previous 40 years in the processing and preparation of foods. This is being done less and less in the home and more and more by food manufacturers, and has resulted in major changes in our diets. Since during this period life expectation at all ages, from the very young to the very old, has increased, these changes cannot be other than beneficial. But there is still an unacceptably large number of premature deaths from degenerative disorders, notably from cardiovascular diseases and cancer, which may be attributed at least in part to diet. In particular, excess intake of saturated fat, sugar, salt or food additives may contribute to some of these early deaths.

Each of these dietary factors was discussed by the speakers. It is beyond dispute that each of these factors may have adverse effects on health and may be partly responsible for early mortality, if the excess or deficiency is very large in a small number of people with metabolic abnormalities arising from their genetic constitutions. What is disputed is whether they have any effect on the health and mortality of the general population whose diets are made up of varying mixtures of foods now readily available. Those who consider that they do, advocate extensive programmes of detailed nutritional education and of government controls of food production and processing by regulations affecting the composition of foods that farmers may produce and manufacturers may sell. These measures have considerable support. However, there are also many who think that the scientific evidence does not justify many of the teachings or policies recommended. Any benefits that might accrue would be offset by giving up the enjoyment of traditional foods that we like such as whole milk, bacon and eggs and many others; there is also the danger of creating a multitude of food neurotics; some of the changes may turn out to have hitherto unrecognized adverse effects on health.

I enjoyed the conference and listening to the debates, and recommend this book strongly. But I warn readers that it raises far more questions than it answers. Even after a careful study of the papers and discussions, they will still have ‘unhappiness of knowledge’, ‘sturdy doubts’ and perhaps ‘boisterous objections’.

R. Passmore

Introduction

The Pfizer Foundation and the Post-Graduate Board of the Faculty of Medicine, University of Edinburgh, asked us to organize the Nutrition Conference as part of the Edinburgh International Festival of Science and Technology. This Festival of Science is intended to complement the widely acclaimed Edinburgh Festival which takes place in late summer.

The remit we were given was to discuss nutrition in the next decade. What should the public be told? There has been much discussion with expert groups as to what food we should be eating, yet the general public is bemused by much of what is being suggested.

This book extends the presentation given at this meeting. The format was that two experts were asked to speak on the major issues of cholesterol and fat, sugar, salt and fibre. The issue was then discussed by them and the audience. Invective was neither looked for nor received. Instead of a similar voice creating a wilderness we believe that a wide discussion of the scientific issues was achieved. Important but less contentious issues such as national food policies, the infant origin of common diseases, preservatives and the food manufacturers' role and problems were also discussed.

The meeting was completed by a consideration of recommendations for the diet of the young, middle aged and elderly.

This volume is not intended as a textbook, nor as the record of a meeting. It is intended to help nutritionists and those involved in the provision of nutrition to the public to make up their own minds on these important issues. The views that are expressed should enable the reader to develop working principles in practical nutrition. In discussion sections, participants are recorded by surname. A full list of names and addresses appears in the Appendix.

We have enjoyed working on this enterprise. We have been helped and supported by contributors, by Dr Stevens of the Pfizer Foundation, Professor Kendall, Dean of the Faculty of Medicine, University of Edinburgh, Dr Muir and Sir James Fraser, Post-Graduate Deans of the Faculty of Medicine, University of Edinburgh. The venue was the Royal College of Physicians of Edinburgh, whose hospitality was so important, and we are grateful to Professor John Richmond, President of the Royal College of Physicians for his presence and hospitality. We would also like to thank Anne Jenkinson, without whose tireless efforts and enterprise this would not have occurred.

Most of all, the audience provided an ambience which was such that a good array of views was achieved.

Martin Eastwood
Christine Edwards
Doreen Parry

1 National food policies

W.P.T. James and A. Ralph

1.1 INTRODUCTION

Most of this volume will be concerned about the new era of science and nutrition and will cover debates on the issues relating to different nutrients. This contribution, however, does not deal with science but with national nutrition policy. When considering policy we sometimes deal only tangentially with science as an attempt is made to integrate an array of different data with the practical issues of implementation. In order to change governmental approaches to society based on the evidence we have, it is essential to make a judgement that any change proposed is going to do more good than harm. Scientists and others may delight in finding fault with a policy, either because they disagree with the scientific issue, perceive an inconsistency in the policy or fail to understand the basis for the policy because they were not part of the group evolving the new approach. If scientists wish to become involved in nutrition policy-making then its uncertainties are very different from scientific research. It is almost inevitable that there will be inconsistencies and problems in developing a particular policy.

Policy-making is a completely different exercise from that relating to health education, which imposes a completely different set of demands. Health education is a rudimentary science of which most of us know practically nothing.

The current British nutrition policy stems from the work of Boyd Orr who established the Rowett Research Institute in 1913. Boyd Orr was trained as a physician but established an agricultural institute because interest and money supported this emphasis in nutrition at the time. His small group of scientists became involved in animal husbandry and came to realize that there were many parallels with the human condition. A series of studies including fairly crude epidemiology and some simple feeding trials led to the conclusion that there was something wrong with the diet of British people. These trials had shown that milk in particular led to a spurt in height growth and surveys showed that children from poor families received little milk and were short and frequently sick. Boyd Orr (1946) related household income to the diet and health of the children and adults and, on the basis of these simple associations, backed up by simple feeding trials on children, he proposed a health nutrition policy for Britain.

These studies of Boyd Orr took place in the 1920s and 1930s but society had been concerned about the nation's health for about 25 years, since the time when young men were being recruited for the Boer War. Physicians monitoring the status of these young potential soldiers concluded that many were unfit to be recruited as infantrymen because they were short, thin and sick. This led to a major societal concern about whether Britain had sickly children growing up into decrepit manhood. There was also concern because at that time the working classes had many more children so a debate emerged as to whether the genetic stock of the British was deteriorating so that we would end up as a puny race unable to govern our colonies. So when Boyd Orr claimed that poor physique was not inevitable and that if children were well fed they would grow better, this struck a chord in those demanding societal change to improve our health.

However, Boyd Orr was not popular when he proposed an input of money for school meals and school milk. But being a canny Scot he recognized the value of milk at a time when farmers were unable to sell all the milk that they produced. Boyd Orr liaised with the Milk Marketing Board and personally began a campaign to try to persuade charitable groups to introduce school milk on a voluntary paying basis, with supplements from the various charitable institutions. This established the principle of need but it was Churchill, at the beginning of the Second World War, who recognized that there was a potential crisis in food supplies and allowed Boyd Orr and others to form a nutrition policy, much of which persists to this day.

1.2 WORLD WAR II

The problem during the Second World War was that only about 30% of our food needs were produced at home so 70% of our food was at the mercy of German submarines as Allied ships tried to bring food from the colonies and across the Atlantic from the United States. The culture was geared to the concept of poor, inadequately-fed children faced with desperately short supplies of food. If the country could provide good food for the people then the health of both children and adults would improve. This would benefit war-time performance and morale. Rationing based on the supposed true needs of individuals during war time was therefore essential and relied on the surveys that Boyd Orr and others had conducted. Boyd Orr and fellow nutritionists propagated the idea that the British needed to produce as much milk, butter, meat and good bread as possible so that these foods could be distributed in subsidized form to the poor of Britain. Food compositional standards had to be met but nutrition education was also very important.

We find today in many government departments food and agricultural policies which were set up at that time. Different ministries operate their own systems unrelated to each other and all based on concepts which emanated from 50 years ago. It seems to take about 30 years to achieve meaningful change in society following the initial warning signals that come from the science of the role of diet in disease.

1.3 POST-WAR DEVELOPMENTS

After the war it was recognized that we had been remarkably successful in our food policies in Britain. Children had grown well, adults had been active and effective with no evidence of deficiency diseases increasing in their prevalence. We therefore seemed to know everything we needed to know about human nutrition. There was a massive shift of thinking in the Medical Research Council which had been a major backer of nutritional research since it was first established. Scientists such as Hans Krebs went from their nutritional studies, for example of vitamin C, back to their first love of biochemistry. A national policy was established that the 30% figure for home-produced food was totally inadequate. Should we face another war, with the Soviet Union for instance, we would need to grow more food in Britain to counter a naval threat to food supplies. Nutritional research in humans was therefore transferred to Jamaica with its Tropical Metabolism Research Unit working on so-called protein deficiency and malnutrition, to Uganda and to India; Britain ended up by 1970 with one Medical Research Council Nutrition Unit, the Dunn in Cambridge.

Agricultural institutes such as the Rowett expanded enormously post-war and several institutes concentrated almost exclusively on animal production. The priorities were to produce more and cheaper food so that the poor could afford it. The more plentiful the food supply and the more intensive the production, the better. Since the Second World War, agriculture has seen one of the most spectacular science-led revolutions in the Western world and agricultural research is probably the one area where publicly-funded science had contributed most to the industrial base of Britain. We now produce not 30% but 80% of our food in the UK as the whole production system was transformed with steadily cheaper food. Meat became plentiful with highly valued beef, veal and lamb becoming readily available from home supplies or from Australasia. later it was realized that pork could be produced more cheaply so there was a marked increase in pork consumption to be followed by a large increase in chicken sales following the development of cheap battery systems of chicken production. Dairy production intensified so there was cheap milk, butter, cream and cheese in abundance. And once food rationing finished in 1952, people could begin to enjoy themselves with luxury foods, confectionery and other special foods previously in short supply. The whole emphasis was on enjoying the increasingly available 'rich' food since deficiency diseases were a feature of the past.

During the 1960s there was a revolution in food processing and retailing with a welter of new products and money to buy them. The growth of supermarket chains allowed everything to be bought at one site, alcohol and cigarettes being added to the food stalls to produce a convenient, pleasurable and profitable enterprise which only now do we recognize as a lethal combination of products.

There is no doubt that post-war children were growing better, infection rates declined, immunization protocols were improving and children were healthier than they had ever been before. However, other concerns were emerging in adults, particularly in relation to cardiovascular diseases and cancers. Their basis was mysterious with Yudkin pointing out that affluence, as indicated by sale of television licences, was a reasonable marker of heart disease but in his view the cause of heart disease was likely to be sugar (Yudkin, 1957). This controversial view was part of a new proposal that diet in some odd way might relate to these diseases of affluence.

1.4 BRITAIN TODAY

We have now reached the situation in Britain where the cheap food policy has allowed the smallest proportion of a household budget of any Western European country to be spent on food. The marked changes in consumption negate the theory of a stable British pattern of food consumption which people imagine to be part of the British fabric of life. In fact we have been altering our diet year by year, decade by decade, since the turn of the century. The post-war emphasis on luxury foods has led to a neglect of potatoes which are regarded as boring, of bread, seen as a 'filler', and of vegetables. There has been a post-war increase in visible fat consumption and an increase in butter consumption until this was partly replaced by cheaper margarines based on imported vegetable oils.

So, for 30 years the British people have apparently enjoyed their food in the knowledge that they were also avoiding protein, vitamin and mineral deficiencies. It is hardly surprising therefore that we are now in the curious position where we have a major public health problem with which many people have not yet come to terms. Figure 1.1 shows how Northern Ireland competes with Scotland for the world's worst record on mortality from coronary heart disease. This figure shows premature coronary heart disease,

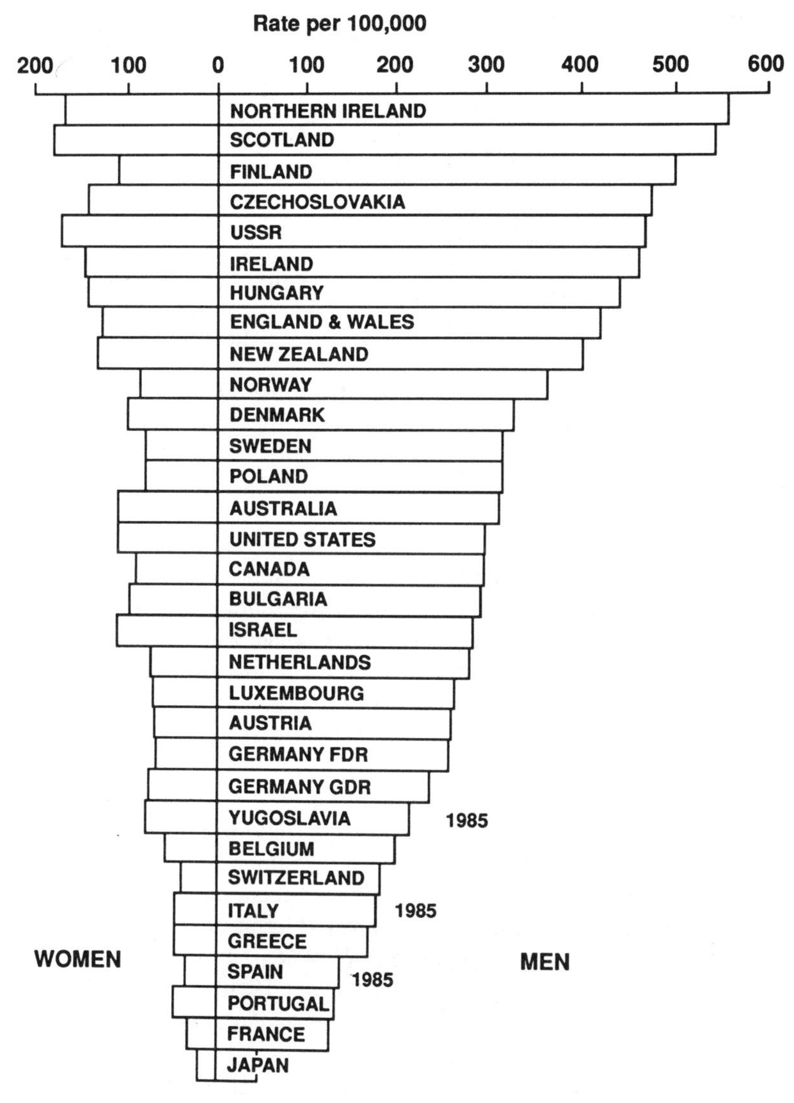

Figure 1.1 Coronary heart disease rates in 32 countries for men and women aged 40–60. Age-standardized rates per 100 000 for 1986, unless otherwise stated.

not the total rates. So why should there be such large differences in coronary heart disease rates across Europe?

Since the 1960s epidemiologists have been highlighting smoking, high blood pressure and high blood cholesterol as the three principal risk factors for heart disease. The links between the then current evidence and policy were eloquently discussed by Marmot (1986). These three risk factors are,

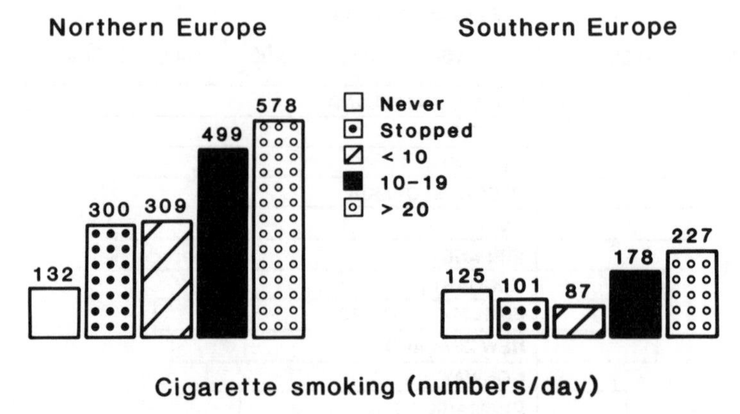

Figure 1.2 Smoking and risk of coronary heart disease in northern and southern Europe.

however, an inadequate explanation of the total range of risk between different groups in Britain and Europe. Alternative theories now abound which attempt to explain the basis of other risk factors. The work of Tunstall-Pedoe (1989) suggests that, in Scottish men, the risk for coronary heart disease of not eating fruit and vegetables is as high as smoking. If we refer to the classic work of Keys (1980), we also find that in Crete, Greece and southern Italy there was an increase in coronary disease risk with increase in smoking rate but the effect was far less than in Northern Europe (Fig. 1.2). So while we accept that smoking is a major risk factor, it is clear that it is affected by other environmental factors and we think diet is one of them. This dietary modulation of the risk of smoking may relate to the antioxidant role of vitamins A, C and E in fruit and vegetables (Duthie *et al.*, 1989).

High blood pressure is an important problem as some recent data from the Oxford Unit clearly illustrate (McMahon *et al.*, 1990). There is a progressive relationship between an individual's diastolic blood pressure and the risk of stroke and coronary heart disease. Thus there is no evidence of a threshold effect with a blood pressure above which the risk begins to increase. However, despite the recognition that excess body weight, alcohol and salt intake have some effect, we do not really know why different groups of adults develop different degrees of high blood pressure.

Total blood cholesterol has long been shown to relate to coronary artery disease risk. Peto (personal communication) has recalculated data from The Seven Countries Study (Keys, 1980) to show a remarkable concordance between total cholesterol and coronary artery disease on a cross-cultural basis (Fig. 1.3) but the men of Crete prove to be an exception with a very low incidence of new coronary heart disease despite cholesterol levels of

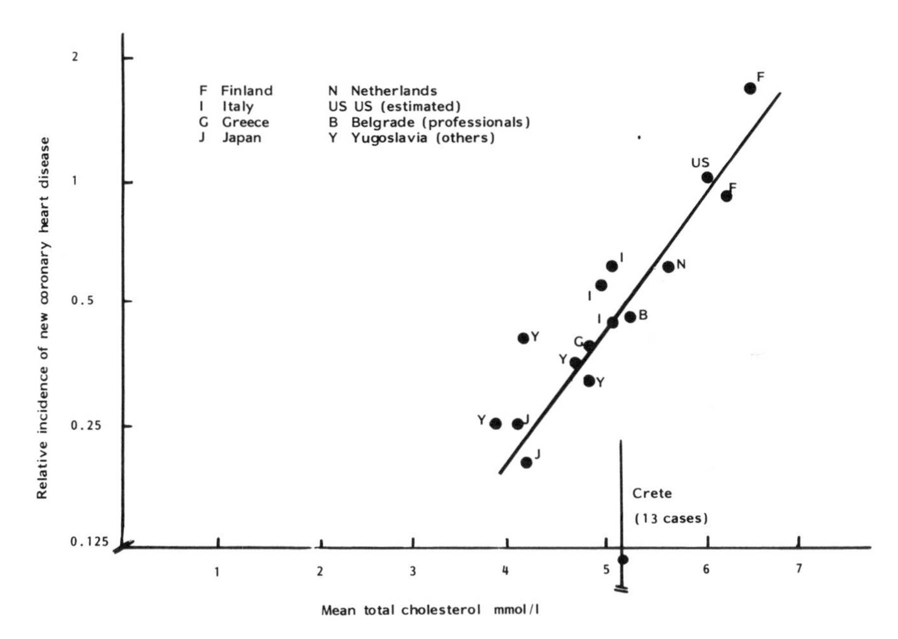

Figure 1.3 Serum cholesterol level and incidence of coronary heart disease in eight countries (Peto – recalculated from Key's Seven Country Study (1980) – personal communication)

5.2 mmol/l. The data from the MRFIT study (Martin *et al.*, 1986) show that there is a relationship between serum cholesterol and death rate within one society.

So if we consider the classic evidence, the clinical studies, animal experiments and the newer intervention trials, we should be able to come up with some policy statements which propose measures to avoid smoking and to limit risks in blood pressure and serum cholesterol. However, the situation in Britain today gives cause for concern. If we take the proposals of the US National Institute of Health as a starting point (European proposals are similar) they suggest a cut-off point of 5.2 mmol/l cholesterol. Levels below this are acceptable but those with cholesterol concentrations between 5.2 and 6.5 mmol/l should have immediate dietary advice and be assessed yearly, whereas those with levels persisting above 6.5 mmol/l should have lipoprotein analysis and, if the LDL fraction is greater than 1.6, then drug therapy should follow. Data for the middle-aged in Scotland (Tunstall-Pedoe *et al.*, 1989) would suggest that 40% of men should be going to a lipid clinic and 80% of them immediately require a new diet. Such management would thus promptly overwhelm the medical services. Alternatively we can tackle this problem by

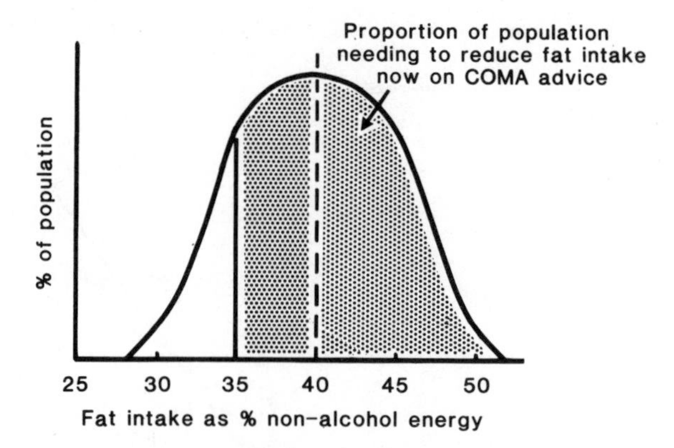

Figure 1.4 Percentage of the population needing to reduce fat intake. Data based on current food intake studies of adults and children in Britain.

taking dietary criteria for change. On this basis the COMA (1984) (Committee on Medical Aspects of Food Policy) recommendations to reduce total fat intake to below 35% will affect 75% of children, adolescents and adults in Britain today (Fig. 1.4). Thus even modest proposals suggest that a large majority of British people have an inappropriate diet.

In the prevention of heart disease one considers how best to deal with defined risk factors. However, where cancer is to be prevented it is no longer possible to operate on specific risk factors. On the basis of a variety of different studies a series of dietary components have been related to a number of different cancers as shown in Table 1.1 (Surgeon General, 1988), but the question is whether to operate on the basis of these associations and suggest dietary changes. Similarly, with the problem of overweight and obesity, it is difficult to know what specific risk factors other than family history to target when dealing with a problem involving 50% of the middle-aged population. Obesity as such is itself a serious risk factor for morbidity and mortality as shown by Manson *et al.* (1990). Therefore expert committees on cancer and obesity have proposed preventive measures. For obesity it is difficult to fault the suggestion that total fat intake is conducive to weight gain. Figure 1.5 shows rather crude data from Brazil which suggest that as the fat content of the diet goes up so does the average body mass index. With many associations between disease and lifestyle, the problem is how to discern the links in a multi-factorial process and then develop clear recommendations for prevention.

Expert reports, including those from the World Health Organization, have developed a series of dietary recommendations about society as a whole,

Table 1.1 Associations between selected dietary components and cancer (adapted and extended from US Surgeon General's Report, 1988)

Selected cancer sites	*Fat*	*Body weight*	*Fibre*	*Fruits and vegetables*	*Alcohol*	*Smoked, salted and pickled foods*
Lung				−		
Breast	+	+			−+/−	
Colon	++		−	−−		
Prostate	++					
Bladder				−		
Rectum	+				+	
Endometrium		++				
Oral cavity				−	+*	
Stomach				−−		++
Cervix				−		
Oesophagus					++*	+

Key + = Positive association; increased intake with increased cancer
− = Negative association; increased intake with decreased cancer
*Synergistic with smoking

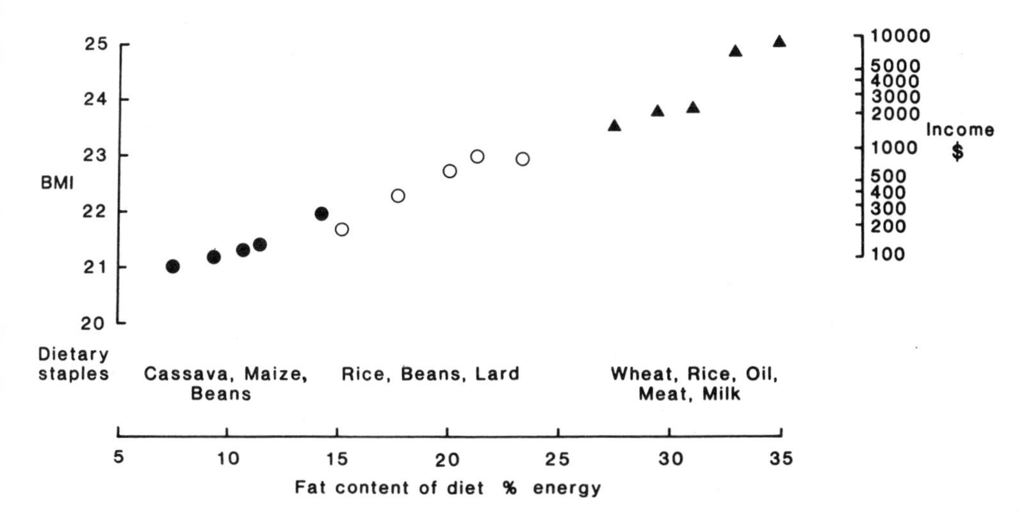

Figure 1.5 Relationship between food intake, income and BMI (body mass index) in Brazil.

aimed at reducing the prevalence of disease. The values for saturated fat, total fat, protein, complex carbohydrates and sugar are given in Table 1.2. Britain is classified with the northern strip of European countries as having a diet that is so far from these goals that the European region of WHO suggested short-term Northern European goals in an attempt to cope with what would

Table 1.2 Intermediate and ultimate nutrient goals for Europe

	Intermediate goals		*Ultimate goals*
	General population	*Cardiovascular high-risk group*	
Percentage of total energy[a] derived from			
complex carbohydrates[b]	>40	>45	>45–55
protein	12–13	12–13	12–13
sugar	10	10	10
total fat	35	30	20–30
saturated fat	15	10	10
P:S ratio[c]	≧0.5	≦1.0	≦1.0
Dietary fibre (g/day)[d]	30	>30	>30
Salt (g/day)	7–8	5	5
Cholesterol (mg/4.18 MJ)	—	<100	<100
Water fluoride (mg/l)	0.7–1.2	0.7–1.2	0.7–1.2

[a] All the values given refer to alcohol-free total energy intakes
[b] The complex carbohydrate figures are implications of the other recommendations
[c] This is the ratio of polyunsaturated to saturated fatty acids
[d] Dietary fibre values are based on analytical methods that measure non-starch polysaccharide and the enzyme-resistant starch produced by food processing or cooking methods

Table 1.3 Changes in fat content of British carcasses in the last decade (Kempster, 1989)

	Separable adipose tissue		*Total lipids*	
	Initial g/kg	*% change*	*Initial g/kg*	*% change*
Cattle	258	2	202	3
Sheep	259	2	235	1
Pig	274	30	223	30

otherwise have meant a huge change in agriculture and food systems. Data from the National Nutritional Surveillance suggest that the major source of saturated fat is animals. This suggests the need for a major change in husbandry, a point suggested by COMA (1984). Yet an assessment of the fat content of British carcasses (Kempster, 1989) suggests that British agriculture has failed to adapt to consumer needs, with the exception of pig rearing where there has been a 40% reduction in the amount of back-fat produced; the changes in sheep and cattle rearing have been negligible (Table 1.3). Unfortunately we have an agricultural policy, a distribution network and farm

Table 1.4 Adult sugar intakes in Cambridge (recalculated from COMA Report on Sugar, 1989)

	Women (gd^{-1})		*Men* (gd^{-1})	
	Extrinsic	*Intrinsic**	*Extrinsic*	*Intrinsic**
Table sugar	27	—	63	—
Biscuits, cakes	10	4	11	5
Milk, butter, cheese and yogurt	2	17	2	18
Totals	62	40	107	43
Extrinsic total as %	61		71	
% hidden extrinsic sugar	56		41	

* Intrinsic includes milk sugars

production all geared to producing as much meat as possible, with little concern for the amount of fat in the carcass. This means that a vast amount of excess fat is going into the food chain as cheap food fillers in sausages, pies, pâtés and elsewhere.

Sugar consumption was examined in the recent COMA Report on Sugar (1989). The latest data (Table 1.4) show that well over half the sugar intake in women and 40% in men is extrinsic. Much of that added sugar is actually hidden in food products. Similarly, with salt, only about 15% of our sodium intake comes from discretionary use in cooking or on the table, the rest being incorporated in manufactured foods (James *et al.*, 1987). So we have a fundamental problem because the whole basis on which consumers can respond to health education messages is flawed: we cannot make informed choices because we are unable to find out what is in our food. Surveys of consumer attitudes to food labelling suggest that people are ignorant and confused about nutrition messages and are misinterpreting nutrition labelling. One can defy any nutritionist with a scientific training, without a microcomputer, to go round a supermarket and actually choose an appropriate diet based on what is recommended in COMA. It is impossible because a coherent approach to nutrition education and food policy is not in place and has not been since the Second World War. Thus major changes in the food chain and major new developments in our understanding of health have not been integrated effectively as part of a new national policy. Some groups such as the Coronary Prevention Group and MAFF (Ministry of Agriculture, Fisheries and Food) are developing new systems of labelling which have yet to be put into practice. On a positive note, we are now beginning to see a downturn in saturated fat intakes in Britain (Fig. 1.6) and in Northern Europe (James *et al.*, 1988) but this is about 20 years behind what has been happening in North America (Stephen and Wald, 1990) and Australasia.

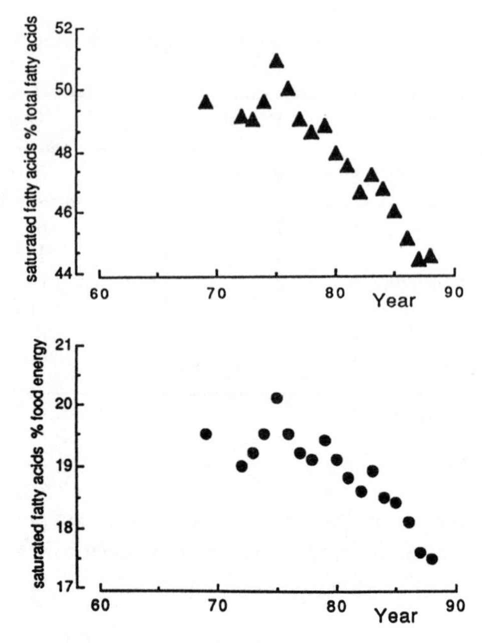

Figure 1.6 Changes in intake of saturated fatty acids in Great Britain 1969–88 (National Food Survey).

Finally, when dealing with policy on a national level we must recognize that we have huge industrial resources in dairying, the bread and confectionery industries which have developed on concepts of diet which are 50 years old and take little or no account of our current knowledge of the science of nutrition. We hope the 1990s will see some changes in policy which pay attention to an overwhelming body of evidence which suggests that our expensively subsidized agricultural processing and distribution system is geared to the wrong priorities.

REFERENCES

Boyd Orr, J. (1946) *Food, Health and Income*, Macmillan, London.

Committee on Medical Aspects of Food Policy (1984) *Diet and Cardiovascular Disease, Report on Health and Social Subjects No. 28*, HMSO, London.

Committee on Medical Aspects of Food Policy (1989) *Dietary Sugars and Human Disease. Report of the Panel on Dietary Sugars*, HMSO, London.

Duthie, G.G., Wahle, K.W.J. and James, W.P.T. (1989) Oxidants, antioxidants and cardiovascular disease. *Nutr. Res. Rev.*, 2, 51–62.

James, W.P.T., Ferro-Luzzi, A., Isaksson, B. and Szostak, W.B. (1988) *Healthy Nutrition: Preventing Nutrition-related Diseases in Europe*, WHO Regional Publications, European Series, No. 24, Copenhagen.

James, W.P.T., Ralph, A. and Sanchez-Castillo, C.P. (1987) The dominance of salt in manufactured food in the sodium intake of affluent societies. *Lancet*, **i**, 426–8.

Kempster, A.J. (1989) Carcass and meat quality research to meet market needs. *Anim. Prod.*, **48**, 483–96.

Keys, A. (1980) *Seven Countries: A Multivariate Analysis of Death and Coronary Heart Disease*, Harvard University Press, London.

McMahon, S., Peto, R., Cutler, J., *et al.* (1990) Blood pressure, stroke and coronary heart disease. *Lancet*, **335**, 765–74.

Manson, J.A., Colditz, G.A., Stampher, M.J., *et al.*, (1990) A prospective study of obesity and risk of Coronary Heart Disease in women. *N Engl. J. Med.*, **322**, 882–9.

Marmot, M.G. (1986) Epidemiology and the art of the soluble. *Lancet*, **i**, 897–900.

Martin, M.J., Hulley, S.B., Browne, W.S., *et al.* (1986) Serum cholesterol, blood pressure and mortality: implications from a cohort of 361 662 men. *Lancet*, **ii**, 933–6.

Stephen, A.M. and Wald, N.J. (1990) Trends in individual consumption of dietary fat in the United States 1920–1984. *Am J. Clin. Nutr.*(in press), **52**, 457–69.

Surgeon General (1988) *Report on Nutrition and Health*, US Department of Health and Human Services, Public Health Service, Publication No. 88-50210.

Tunstall-Pedoe, H. (1989) Coronary risk factor and lifestyle variation across Scotland: results from the Scottish Heart Health Study. *Scott. Med. J.*, **34**, 556–60.

Tunstall-Pedoe, H., Smith W.C.S. and Tavendale, R. (1989) How-often-that-high graphs of serum cholesterol. Findings from the Scottish Heart Health and Scottish MONICA studies. *Lancet*, **i**, 540–2.

Yudkin, J. (1957) Diet and coronary thrombosis: hypothesis and fact. *Lancet*, **ii**, 155–62.

DISCUSSION

Kritchevsky, Philadelphia Just to compound the confusion, you show a slide from the US nurses study indicating the steady increase in coronary disease on a percentages basis. The increase in fat was taken from a previous study which has shown that the calories provided by fat in a diet had absolutely no effect on breast cancers. I think we have to look at these as diseases which increase as the population becomes older. We do have to deal with them but I don't think there are any simple solutions. In addition there are now substantial animal data which shows that total calories regardless of source are more important *vis à vis* production of tumours than is dietary fat, carbohydrate or protein.

James, Aberdeen Thank you, I glossed over the issue of whether fat does really cause breast cancer. I am familiar with and have had much debate with Walt Willet on the US nurses study, which of course came up with the point that there seems to be no relationship between fat intake and breast cancer whereas there was a surprising relationship, as they thought, with alcohol consumption and breast cancer. I am worried about this concept of calories which emanates from animal studies because it is perfectly possible to manipulate the energy intake of animals and slow tumour growth. The question is how to apply this to man and what to do in practice. It is remarkably difficult to semi-starve children or indeed adults. We have an amazing physiological control of food intake. I personally do not agree with the concept that energy as such is the key component in humans because most of the statistical analyses that have been undertaken on the energy/fat relationship have not taken into account the concordant relationship between energy intake and body weight. Once you start stripping out the energy factor on a per kg basis as a way of simplification then one does not get the same relationship between energy intake and cancer that one might expect from the animal data.

Moynahan, London I would like to follow up what has just been said. I think the most important factor is the control of infection. It is the first time in human history that the majority of the population has survived beyond perhaps the age of 40 and so we may be seeing coronary disease or tumours or whatever it is emerging as a result of longevity from control of infection. In fact we are very unlike wild animals in that respect. We are in control of our population. At the moment in the UK one in five of the population is now over 60 and, at the rate we are going with the decline in birth rate, before the end of the decade it will be as high as one in three. I noticed that in the United States if you compare the fat changes in the coronary vessels in American servicemen killed in the Korean war and later in Vietnam you will find that there has been a significant decrease in the fat which has been found in coronaries. I don't know how it has occurred and what it has been attributed to, whether the Americans are in fact eating less fat.

Wilson Nicol, Reading There are two points I would like to make: one, it is very difficult for me to believe that fat is a cheap filler because it costs several times more than carbohydrates and two, I find it rather strange that you refer to 56% of sugar being hidden when in fact 100% of flour and most other ingredients are also hidden by your same definition.

James Fat is a cheap filler. Of course, it is much more expensive to produce agriculturally and that is the nonsense because consumers do not want it and the farmers have to expend far more feed in producing it. But it does become cheap once it cannot be put overtly into the food chain. In other societies where there is an even higher premium on lean meat, for example, they actually regard the fat content that is stripped off as totally superfluous. It is only in that operational marketing sense that it is a cheap filler.

When it comes to sugar, why bother with sugar rather than flour being hidden? Simply because the advice of government committees is that you should attempt to reduce your sugar intake particularly between meals and particularly if you are overweight. How do you do that? You must know where the sugar comes from. That is my only point in relation to labelling. I totally agree, infection rates are dropping. The evidence on the pathology of coronary heart disease in America is absolutely true; Sinclair suggested that if you take actual data on the fatty acid profile of fat biopsies, taken for a variety of different studies, you can show a beautiful inverse relationship between the linoleic acid content of the adipose tissue and the coronary disease of that particularly community, implying that it is the type of fatty acid which is important in explaining the change, but this of course is not proven.

Morgan, Glasgow The good news, in yesterday's local paper in Glasgow, is that the meat producers are producing lean beef. The bad news is the decrease in healthy food consumption in Italy. The question is about alcohol. The slide showed 35% fat intake not including alcohol. What proportion does alcohol play in the total fat intake?

James The advice on alcohol is to bring the total average national intake below about 4%. As far as the relationship between alcohol and dietary fat content is concerned, the fat content is expressed on a non-alcohol basis. If you are going to try and interpret this for children, the assumption, a rather foolish one these days, is that children are not consuming alcohol. If you monitor people who are on a low fat diet in Britain, very frequently you find that if you express the fat on the basis of total energy the reason why they are on a low fat diet is because they are either on a high alcohol or a high sugar diet, implying that we have an even bigger problem as we try to shift dietary patterns.

Eastwood, Edinburgh What you are showing are changing attitudes to nutrition over the last fifty years, and each time they change, nutrition becomes denegrated. The population that was brought up with 'milk and cheese are good for you' are now told that they are not good for you, and we were told recently that oats were good for us and now we are told that they are not. Each time this happens nutrition becomes a very poor science and is diminished in worth. I would like to know how do you cope with change in nutritional policies against a background where there is an undermining of previous advice?

James I think that this is a problem much exaggerated by the media because the actual swing in nutritional thinking is amazingly slow. One also tends to get crude statements, for example 'that meat is bad for you'. Curiously enough, it is meat fat that is bad for one. The COMA discussions failed to recognize the implications for nutrition policy, for education and for agriculture of not making the distinction between the fat content of meat and the protein content of meat. I think we have to be more sophisticated in developing health education messages. We also have to accept and cope honestly and openly where shifts in thinking occur when they do occur. It is a slow evolution in thought.

MacDonald, London You said that nutrition education was a minefield. What proposals do you have to get through this minefield?

James I think we have to get to grips with the whole mechanism whereby nutrition education can be developed. Generic statements such as 'eat a balanced diet' or 'eat less of this' are totally useless. The problem is how to cope on an individual level with people of different ages, different heights and sizes, different individual requirements and at the same time get these concepts across. It is a huge problem and I do not think as yet it is one that we have tackled in a proper practical way.

Passmore, Edinburgh Could you amplify about the big difference in coronary death rate between Scotland and France. It has puzzled me for a long time. The overall death rates differ very little. Other diseases must make up the difference. Has it any thing to do with the larger intake in alcohol and garlic in France?

James Garlic probably works through allyl sulphides which affect the plasma lipids, but this is unlikely to be very important. My particular theory is that an antioxidant effect of plant foods could explain the difference. When it comes to death rates, I think that when you are looking at total death rates or longevity, you are looking for fractional changes. I was talking about premature death rates, not implying that we had a recipe for immortality.

2 Infant origins of common diseases

D.J.P. Barker

2.1 INTRODUCTION

In this paper I will present evidence which suggests that the nutrition of an infant and of its mother are much more important in the causation of cardiovascular disease than we have previously supposed. The starting point is the well known uneven distribution of deaths from ischaemic heart disease in England and Wales. Essentially the pattern is of low rates right through the South of England including London, which is a point of considerable importance, and high rates in two kinds of area in the north – industrial towns and in some rural areas which have the poorest agricultural soil in the country (Gardner *et al.*, 1969, 1984). The pattern of stroke mortality is similar to that of ischaemic heart disease. Furthermore this pattern is specific to cardiovascular disease: it is not for example, shared by chronic bronchitis.

Although smoking and adult diet affect the cardiovascular risks of an individual, and the differences in risk between populations, they do not offer a sufficient explanation of these large differences in death rates between one part of Britain and another. The diet of people in different areas in Britain is in fact remarkably similar (Cade *et al.*, 1988). It seems logical that if we cannot explain the distribution of cardiovascular disease by looking at the adult environment perhaps there are answers to be obtained by examining what happened to people when they were children.

2.2 GEOGRAPHICAL DISTRIBUTION OF INFANT MORTALITY

The reason we can examine in detail the early life experiences of people in different parts of Britain arises out of what Professor James has already referred to – the concern about the quality of the nation's physique which came to light during recruitment for the Boer war. This led to detailed studies of child health. The maps of infant mortality throughout England and Wales in the early years of this century are strikingly similar to those of

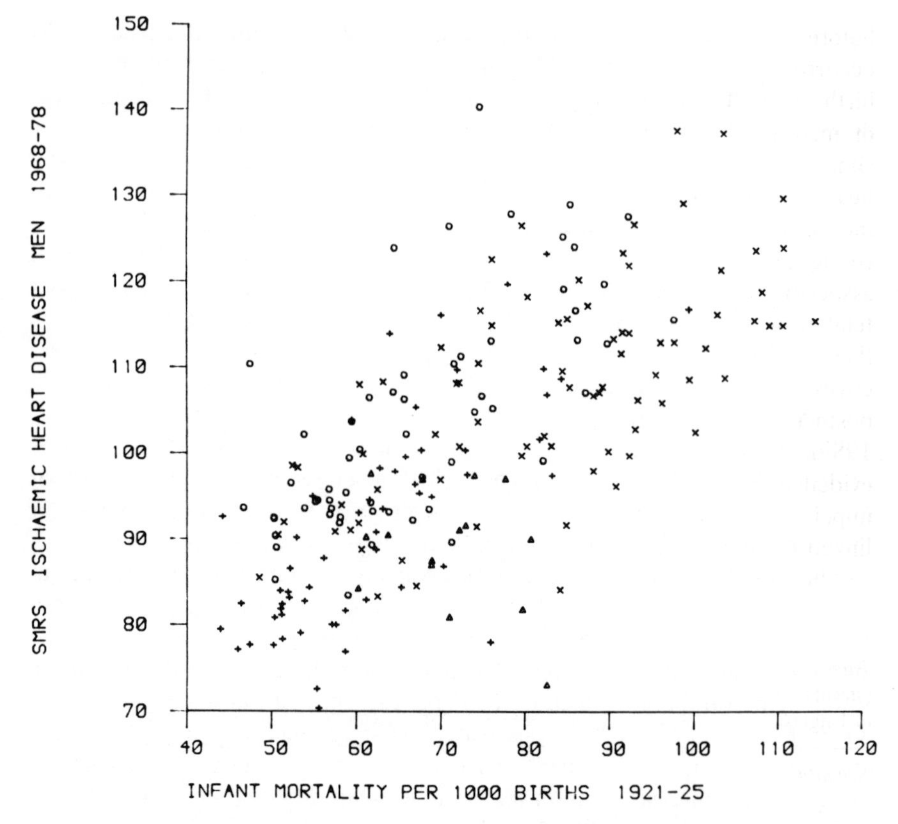

Figure 2.1 Standardized mortality ratios for ischaemic heart disease in men during 1968–78 at the age of 35–74 and infant mortality per thousand births in 1921–5 in the 212 areas of England and Wales (X = county boroughs, △ = London boroughs, ○ = urban areas, + = rural area).

cardiovascular disease today. To put this comparison more formally, Fig. 2.1 shows the relation between infant mortality in 1921–5 and ischaemic heart disease during 1968–78 in the 212 local authority areas into which England and Wales are usually divided (Barker and Osmond, 1986). The relation is strong (correlation coefficient 0.69). It is similarly strong in women. Similar relations are found if infant mortality during earlier time periods is used. The period 1921–5 is presented here because it is the earliest time period during which infant deaths were not only divided into early (neonatal) and late (postneonatal) deaths, but assigned to cause.

2.3 CARDIOVASCULAR DISEASE AND INFANT DEATH

Further analysis shows that the geographical pattern of death rates from cardiovascular disease most closely resembles that of neonatal mortality (death before one month of age) in the past. At that time most neonatal deaths occurred during the first week after birth and were attributed to low birthweight (Local Government Board, 1910). The geographical distribution of maternal mortality, from causes other than puerperal fever, was closely similar to neonatal mortality (Barker and Osmond, 1987a). Poor physique and health of the mothers was clearly implicated as a cause of high maternal mortality, and was partially a result of poor nutrition and impaired growth of young girls (Campbell *et al.*, 1932). There is, therefore, a geographical association between high death rates from cardiovascular disease and poor fetal growth and poor maternal physique and health in the past. In addition to these associations, which indicate the importance of the intra-uterine environment, the distribution of ischaemic heart disease is also related to postneonatal mortality (death from one month to one year) (Barker *et al.*, 1989). Most of these later infant deaths were due to infection. There is evidence that in places where infant death rates were high, infant growth was impaired. Ischaemic heart disease, but not stroke, is therefore geographically linked to an adverse environment in infancy as well as in fetal life.

The specificity of the geographical relation of cardiovascular disease to

Table 2.1 Death rates from stroke, ischaemic heart disease and chronic bronchitis (standardized mortality ratios ages 35–74, both sexes, 1968–78) in the 212 areas of England and Wales grouped by neonatal and postneonatal mortality (1911–25)

Neonatal mortality	*Postneonatal mortality*				
	lowest 1	*2*	*3*	*4*	*5 highest*
			Stroke		
1 lowest	85	81	79	78	79
2	86	90	98	74	76
3	102	100	104	104	104
4	—	108	110	115	117
5 highest	124	—	121	123	117
			Ischaemic heart disease		
1 lowest	84	89	91	88	98
2	85	93	95	88	91
3	86	94	99	106	113
4	—	98	109	111	115
5 highest	83	—	114	119	116
			Chronic bronchitis		
1 lowest	67	78	106	115	161
2	64	84	85	104	126
3	69	65	89	88	151
4	—	91	99	120	142
5 highest	41	—	108	123	144

infant death around 70 years ago is shown in Table 2.1. To construct this table the 212 areas of England and Wales were ordered according to neonatal mortality during 1911–25 and divided into five groups of increasing mortality. Five groups with increasing postneonatal mortality were similarly derived. In this way, current mortality by area could be examined within a grid of neonatal and postneonatal mortality 70 years ago. The pattern is clear. Within any of the five bands of postneonatal mortality, standardized mortality ratios for stroke increased sharply with neonatal mortality. There was no independent trend in stroke mortality with postneonatal mortality. Mortality from ischaemic heart disease had similar but separate trends with neonatal and postneonatal mortality. In contrast, mortality from chronic bronchitis showed a steep increase in postneonatal mortality, but no independent trend with neonatal mortality.

The local authority areas of the country with low neonatal mortality but high postneonatal mortality in the past were mostly in London, and are of particular interest. It seems that London provided a good intra-uterine environment but a poor postnatal one. An explanation is the large-scale migration of girls into London, to work as domestic servants. They came from the rural counties in southern England and were noted at the time to have unusually good health and physique.

2.4 HERTFORDSHIRE INFANT FOLLOW-UP STUDY

Clearly this kind of ecological data is a strong pointer towards further research, but does not itself constitute proof. The way forward is to identify records of people born around 70 years ago, for whom early growth and details of the mother were recorded, and to follow up these people and relate these early measurements to their later health. During the last four years, the Medical Research Council has mounted a nationwide search for records of people born in 1910–30, for whom data of the kind required were recorded. Hertfordshire is one of the places where such data have been found in considerable quantities (Barker *et al.*, 1989). From 1911 onwards the attending midwife was required to notify every birth to the County Medical Officer of Health within 36 hours. Almost all births occurred at home. The name and address of the mother, the date of birth and the birthweight were registered. The local health visitor recorded her observations on a form when she visited the home periodically throughout the first year. After a year the form was returned to the County Health Visitor and data were abstracted onto the register, including weight at one year and whether breast fed from birth, bottle fed or both.

It is therefore possible to follow up people born in Hertfordshire and, if they are alive, to measure their cardiovascular risk factors, and if they are dead

Table 2.2 Standardized mortality ratios for ischaemic heart disease according to weight at one year in 5225 men who were breast fed

*Weight (pounds)**	*Standardized mortality ratios†*
≤18	112 (33)
19–20	81 (71)
21–22	100 (154)
23–24	69 (85)
25–26	61 (40)
≥27	38 (9)
TOTAL	81 (392)

*1 pound = 0.45 Kg
†Numbers of deaths in parentheses

to determine what the cause of death was. The measurements and cause of death may be related to birthweight and weight at one year. Table 2.2 shows the result of the first follow-up of 5225 men who were born during 1911–30 in six districts in the county and who were breast fed. Hertfordshire is a prosperous part of England and rates of ischaemic heart disease are below the national average which, when expressing rates as standardized mortality ratios (SMR), is set as 100. Among men whose weights at one year were 18 lb or less the death rates were around three times greater than those who attained 27 lb or more. This is a strong relation: it spans more than 60 years, and it is graded. There was no similar relation in men who were bottle fed from birth, but the numbers are small.

Both prenatal and postnatal growth are important in determining weight at one year, since few infants with below average birthweights reach the heaviest

Table 2.3 Standardized mortality ratios for ischaemic heart disease according to birthweight and weight at one year in men who were breast fed

*Weight at one year (pounds)**	*Weight at birth (pounds)*			
	Below average (≤7)	*Average (7.5–8.5)*	*Above average (≥9)*	*Total*
Below average (≤21)	100 (80)	100 (77)	58 (17)	93 (174)
Average (22–23)	86 (34)	87 (67)	80 (29)	85 (130)
Above average (≥24)	53 (14)	65 (42)	59 (32)	60 (88)
Total	88 (128)	85 (186)	65 (78)	81 (392)

*1 pound = 0.45 Kg

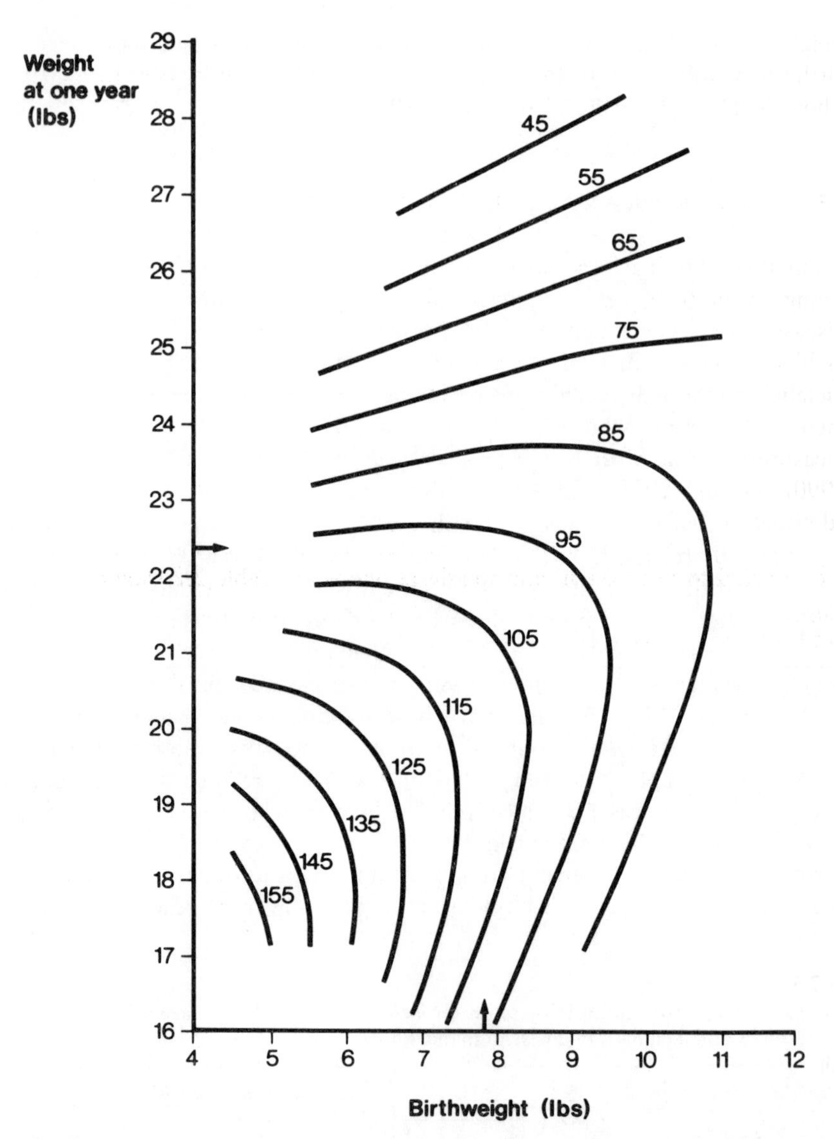

Figure 2.2 The relative risks for ischaemic heart disease in men who were breast fed according to birthweight and weight at 1 year. Arrows indicate average birthweight and average weight at 1 year.

weights at one. The lowest SMRs occurred in men who had above average birthweight or weight at one year (Table 2.3). The highest were in men for whom both weights were below average (Fig. 2.2).

2.5 PRESTON INFANT FOLLOW-UP STUDY

From these findings we can conclude that processes linked to growth and acting in prenatal or early postnatal life strongly influence risk of heart disease. There is evidence that these processes include (1) the determination of blood pressure in fetal life and (2) the long term 'programming' of lipid metabolism through feeding during infancy. In a recent study we traced 449 men and women who were born in a hospital in Preston and whose measurements at birth had been recorded in unusual detail (Barker *et al.*, 1990). During 1935–43 a standardized record was kept for each woman admitted to the labour ward at Sharoe Green Hospital in Preston. The record

Table 2.4 Mean systolic and diastolic pressures (mm Hg) of men and women aged 46 to 54 years according to placental weight and birthweight

*Birthweight (pounds)**	*Placental weight⁺ (pounds)*				
	−1.0	*−1.25*	*−1.5*	*>1.5*	*All*
			Systolic pressure		
≤ 5.5	152	154	153	206	154
	(26)	(13)	(5)	(1)	(45)
− 6.5	147	151	150	166	151
	(16)	(54)	(28)	(8)	(106)
− 7.5	144	148	145	160	149
	(20)	(77)	(45)	(27)	(169)
> 7.5	133	148	147	154	149
	(6)	(27)	(42)	(54)	(129)
All	147	149	147	157	150
	(68)	(171)	(120)	(90)	(449)
			Diastolic pressure		
≤ 5.5	84	87	87	97	86
− 6.5	84	88	85	93	87
− 7.5	84	84	84	90	85
> 7.5	78	85	85	88	86
All	84	86	85	89	86

*1 pound = 0.45 kg

⁺Numbers of people in parentheses

contained details of the mother's previous pregnancies and external measurements of her pelvis. The baby's birthweight, placental weight, length from crown to heel and head circumference were also recorded. Table 2.4 shows the mean systolic and diastolic pressures according to placental weight and birthweight. Blood pressure rose with increasing placental weight but fell with increasing birthweight. Regression analysis showed that mean systolic pressure rose by 15 mm Hg as placental weight increased from $\leq$1 lb to $>$1.5 lb. Systolic pressure fell by 11 mm Hg as birthweight increased from $\leq$5.5 lb to $>$7.5 lb. These relations were independent so that the highest blood pressures occurred in people who had been small babies with large placentas. Higher body mass index and alcohol consumption were also associated with higher blood pressure, but the relations of placental weight and birthweight to blood pressure and hypertension were independent of these influences. These findings show for the first time that the intra-uterine environment has an important effect on blood pressure and hypertension in adults. Our thesis is that discordance between placental and fetal size may lead to circulatory adaptation in the fetus, altered arterial structure in the child and hypertension in the adult.

These findings raise the question of what environmental influences act on the mother and determine discordance between placental and fetal size, leading to high blood pressure. Little is known about this. We suggest that maternal nutrition is one major influence.

Work on baboons has shown that infant feeding may have a permanent effect on the programming of lipid metabolism (Mott, 1986). Little as yet is known about these mechanisms in humans, although studies being carried out by Dr Lucas at the Dunn Unit are beginning to show the important effects of nutrition on the development of premature babies.

In summary, conclusions from geographical analyses in England and Wales are beginning to be borne out by long-term studies of individuals. In Hertford we have learnt that poor prenatal and postnatal growth are strong predictors of risk of death from ischaemic heart disease. In Preston we have learnt that prenatal growth has a major effect on blood pressure in adult life. Examination of people living in Hertfordshire is beginning to show strong relationships between early growth and lipid metabolism, consistent with findings in experimental animals.

2.6 UNEQUAL DEATH RATES IN THREE LANCASHIRE TOWNS

Finally, to return to geographical studies within Britain, the potential of these findings to explain today's inequalities in health can be seen in a study of death rates in three Lancashire towns, Burnley, Nelson and Colne (Barker and Osmond, 1987b). These three towns are situated side by side on the

Table 2.5 Nelson, Colne and Burnley; infant mortality rates per 1000 births from 1896–1925 and infant mortality by period and cause, child mortality and birth rate 1911–13

	Nelson	*Colne*	*Burnley*	*England and Wales*
Infant mortality per 1000 births,				
1896–8	154	170	197	155
1907–10	107	130	171	113
1911–13	87	130	177	111
1921–5	79	109	114	76
Rates from 1911–13				
Infant mortality per 1000 births				
Neonatal	38	37	49	
Postneonatal	49	93	128	
Cause:				
Group of five diseases*	35	33	53	
Bronchitis and pneumonia	17	25	26	
Diarrhoea	16	30	48	
Mortality at age 1–5 years per 1000 survivors at age 1	58	85	96	
Birth rate per 1000 population	18	21	23	

* Premature birth, congenital defects, birth injury, want of breast milk, and atrophy, debility, and marasmus

western slopes of the Pennine hills. Each developed as a cotton-weaving town, and for the 6 miles from the centre of Burnley through Nelson to Colne, there is hardly a break in the line of houses. Mortality in the towns differs considerably. During 1968–78 mortality at all ages and from all causes was 21% above the average in Burnley, 10% above the average in Colne and 4% above in Nelson. Eighty per cent of the excess mortality in Burnley was certified as due to cardiovascular disease, chronic bronchitis or bronchopneumonia.

The close proximity of the towns precludes explaining the large differences in mortality in terms of environmental variables such as rainfall. It is unlikely that there are important differences in medical care. The towns are similar in social class structure, housing and income. The similarity today, however, belies the large differences that formerly existed and which led to large differences in mortality among infants and young children (Table 2.5). These differences in mortality included the health and physique of mothers and in infant-feeding practices. Nelson was the newest of the three towns, developing rapidly from 1870 onwards. Many of the mothers of the present generation of older people were recent migrants from nearby rural areas and not second or third generation industrial workers such as lived in Burnley. More of the women in Nelson, described at the time, as 'sturdier and healthier' than those in Burnley, breast fed their infants and they did so for longer. The birth rate in Nelson was the lowest of the three towns.

2.7 CONCLUSION

In conclusion, there is increasing evidence that maternal nutrition, through its effects on fetal and infant development, has a major effect on the incidence of cardiovascular disease in the next generation. I believe that we will rapidly discover more about the links between early development and adult disease, their strength and the processes which underlie them. If the early promise of this research is fulfilled we will need a new national strategy for preventing cardiovascular disease in Britain. The current strategy is focused on adult lifestyle, the new one will need to address differences in the nutrition and health of mothers and children.

This is a meeting about food policy in the 1990s. One would like to think that in sharp contrast to the 1970s and 1980s our new food policy will be determined not by reliance on outworn ideas, nor by preconception and evangelism, but by science.

REFERENCES

Barker, D.J.P. and Osmond, C. (1986) Infant mortality, childhood nutrition and ischaemic heart disease in England and Wales. *Lancet*, **i**, 1077–81.

Barker, D.J.P. and Osmond, C. (1987a) Death rates from stroke in England and Wales predicted from past maternal mortality. *Br. Med. J.*, **295**, 83–6.

Barker, D.J.P. and Osmond, C. (1987b) Inequalities in health in Britain: specific explanations in three Lancashire towns. *Br. Med. J.*, **294**, 749–52.

Barker, D.J.P., Bull, A.R. Osmond, C. and Simmonds, S.J. (1990) Fetal and placental size and risk of hypertension in adult life. *Br. Med. J.*, **301**, 259–62.

Barker, D.J.P., Osmond, C. and Law, C. (1989) The intra-uterine and early postnatal origins of cardiovascular disease and chronic bronchitis. *J. Epidemiol. Community Health*, **43**, 237–40.

Barker, D.J.P., Winter, P.D., Osmond, C., *et al.* (1989) Weight in infancy and death from ischaemic heart disease. *Lancet*, **ii**, 577–80.

Cade, J.E., Barker, D.J.P., Margetts, B.M. and Morris, J.A. (1988) Diet and inequalities in health in three English towns. *Br. Med. J.*, **296**, 1359–62.

Campbell, J.M., Cameron, D. and Jones, D.M. (1932) *High Maternal Mortality in Certain Areas*. Ministry of Health Reports on Public Health and Medical Subjects, No. 68, HM Stationery Office, London.

Gardner, M.J. Crawford, M.D and Morris, J.N. (1969) Patterns of mortality in middle and early old age in the county boroughs of England and Wales. *Br. J. Prev. Soc. Med.*, **23**, 133–40.

Gardner, M.J., Winter, P.D. and Barker, D.J.P. (1984) *Atlas of Mortality from Selected Diseases in England and Wales 1968–78*, Wiley, Chichester.

Local Government Board (1910) *Thirty-ninth Annual Report 1909–10. Supplement on Infant and Child Mortality*. HMSO, London.

Mott, G.E. (1986) Deferred effects of breastfeeding versus formula feeding on serum lipoprotein concentrations and cholesterol metabolism in baboons, in *The Breastfed Infant: A Model for Performance*. Report of the ninety-first Ross conference on pediatric research (eds L.J. Filer Jr and S.J. Fomon), Ross Laboratories, Columbus, Ohio, pp. 144–9.

DISCUSSION

Eastwood, Edinburgh This is a fascinating long term study. It is always interesting when you have the lifework of a hitherto unknown person being used and their vision being appreciated. Could you start off by talking about why you think that London is curious, because London does not strike a Northerner as being one of the better places to live in, especially with the East End.

Barker No, this is why London is so fascinating because you have the picture painted by Dickens and, in great detail, by Charles Booth in his many volumed report on life and labour in London around the turn of the century. What Booth pointed out and what is very easy to demonstrate is this, London did what no other great city did and that was it constantly renewed itself by massive migration of the young men and women into it from the surrounding counties, particularly the southern and eastern seaboard. Young people came into London in enormous numbers. In 1910 only 60% of women of reproductive age living in London were born there. Most of the rest were born in the southern seaboard and had come into London mainly for domestic service, although there were other employments. So the good health of young women in London is easy to document. As Booth said, the effect of London is a spoiling of people, so in the third generation people are reduced to the level that you would expect from the surroundings they live in.

Garrow, London Two questions. In many surveys of risk factors for coronary heart disease, tallness is a powerful protective factor among adults. Do you think that this is explained by early nutrition or do you think that this is genetic. If it is genetic could that also explain the birthweight, i.e. is the birthweight genetic also? The second question relates to twins. Obviously what you say about birthweight relates to all sorts of social circumstances; however, within each community some children are born as twins and some as singletons. The twins have shared the nutritonal establishment designed for one and therefore tend to be smaller at birth. Is it true or would you expect it to be true that twins are more susceptible to cardiovascular disease?

Barker To take the first point about height, it is quite true that every study that I know of has shown an inverse relationship between height and risk of death from cardiovascular disease. I can not really proceed far with that because height is made up of both prenatal and postnatal components. Your birthweight does predict your height, your height at two years strongly predicts your adult height, so my sense of it is that the relationship does depend on prenatal and very early postnatal components of adult height. We have just finished a survey of 700 men in Hertford. Their height is closely related to their birthweights and weights at one year. I cannot go along with the genetic argument because height changes so quickly when populations are exposed to different circumstances. The Chinese in Hong Kong would be one example. Fetal growth too is highly susceptible to environmental influence. As to twins, this is very interesting and I am sorry that I cannot give you an answer because we have not yet studied them.

Garrow I mean twins as opposed to singletons.

Barker Yes, do you mean, do they die more of heart disease? They do have lower birthweight, they have lower measured intelligence in childhood and I think if you are saying *a priori* you would suspect that they have higher risk of ischaemic heart disease, I think that is a reasonable hypothesis.

Nicoll, Reading What about migration from one part of the country to another? How does that affect your figures?

Barker It would make them worse, would it not? So it makes the strong ecological correlations even more amazing.

Lean, Glasgow In terms of the relationship between low birthweight, low weight at one year and death by cardiovascular or respiratory disease, could this relate to smoking in any way?

Barker Well I think that smoking in women in Hertfordshire in 1920 was not a problem. They did not, so it does not work that way round. There is accumulating evidence that lower respiratory tract infection before the age of two is a major risk factor for chronic obstructive lung disease. All the evidence points to that. The idea that it bears more heavily on smaller infants is certainly reasonable, although short of being proven. Smoking is a major adult risk factor for chronic obstructive lung disease, but it may not be a sufficient factor by itself.

Wharton, Glasgow Could we just tease out the neonatal from the postnatal a bit more? In the association studies, the relationship of adult cardiovascular disease was to neonatal but not to postnatal weight?

Barker Yes, in part to postneonatal, but primarily to neonatal.

Wharton But in the Hertfordshire study, as I understood it, it was really the weight at one year that was the most predictive. That is suggesting more postneonatal events is it not?

Barker Your weight at one year is massively dependent in Hertfordshire on your birthweight. It is almost impossible to reach the heaviest weights at one year if you are not of above average birthweight. Now, in Hertfordshire, you were in a place where postnatal life was very good and there was no substantial overcrowding. Infant mortality from bronchitis and from diarrhoea was low, so you are seeing a situation there, I think, where nutrition was the major determinant of growth up to one year. If you did the same study in Preston it might be very different.

Wharton I wondered whether if you took weight at one year and subtracted birthweight, would that be more predictive as it would summarize postnatal nutrition?

Barker The trouble is that the weight at one year is spread out much more. It is statistically unsatisfactory to do that. It just gives a ghost answer to the much clearer answers you get by the Cox's model which I showed. You cannot take gain in weight independently of the initial weight.

Edwards, Edinburgh If the birthweight is increasing now do you predict a decrease in cardiovascular disease in 60–70 years time?

Barker Well I think that is a clear prediction, not from increases in birthweight, but from the driving force of improvements in maternal nutrition. My prediction is that ischaemic heart disease rates will continue to fall in this country providing the adult environment is held constant.

Wharton, Glasgow There has been only a minimal increase in birthweight in this country since 1945. The only group in which it is increasing is the immigrant population and goes back to what you were saying about genetic factors not being that important in this, it is more the environment.

Barker Do you not think that the problem with birthweight is that it is such a crude assessment of a fetus which does not take into account length, head size, etc. These relative measurements are important in studies of hypertension and are beginning to show predictions in a way that crude birthweight does not. Also the presumption that birthweight is solely an expression of maternal nutrition is flawed because one of the broad principles of nature is that women do not grow babies that are too big to escape by the normal route. There is a constraint of fetal growth towards the end of pregnancy which is made up afterwards.

Palmer, Cambridge I wondered if you have any comparison with different infant societies in Africa or Asia?

Barker We have done a study of 700 children in the Gambia and have found an interesting inverse relationship between children's blood pressure and maternal weight gain. I do not think there are any data in the Third World that would enable you to follow people from birth to death as you can in Hertfordshire.

Kirk, Tayside The mortality data that you presented were for men and I wondered if there were similar patterns for women.

Barker I do not yet know. Women change their names which makes them very difficult to trace, but we have traced around 5000 women addressing just that point.

Armour, London If an individual knows he was a bouncing baby, does he not have to worry about his dietary intake?

Barker A big breast-fed baby was a long baby not a fat one. Yes, I think if you are tall, that is good news and if you are short then you had better get on a bit with your life.

Jackson, Southampton David, the modern breast-fed babies, if we are to believe what we are told, grow well in the first 6 months of life but drop a few percentiles during the second 6 months of life. At one year of age they are below the NHS standards. Would that in any way confound our interpretation of what is likely to happen when they become adults?

Barker All is speculation until we know more about the basic mechanisms which underline these long-term relations.

Kritchevsky I think it should also be pointed out that there are now data emerging that tall people are more susceptible to cancer.

Barker Yes, this is very interesting. Can I just qualify that. If, you look at the counties of England and Wales and you look at the average height of the adult population you get the following: the counties with the shortest average height have the highest mortality from cardiovascular disease, the counties with the tallest heights, and this is eastern sub-seaboard counties, have the highest mortality from three cancers – breast cancer, ovarian cancer and prostatic cancer only. These are the three hormone-dependent cancers.

PART ONE

Cholesterol and Fat

3 Atherosclerosis in perspective: the pathophysiology of human cholesterol metabolism

J. Shepherd and C.J. Packard

3.1 INTRODUCTION

All animal cells share a common need for cholesterol which they obtain either by endogenous synthesis or by assimilation from the diet. The steroid nucleus plays both structural and metabolic roles. In cell membranes it is considered to act as a regulator of the micro-environment, maintaining a fluidity appropriate to the normal operation of membrane-linked enzyme systems and transport proteins. In specialized tissues such as the adrenal, gonad and liver it undergoes a variety of oxidation steps which increase its polarity and permit its direct transit through the aqueous fluids of the body. The parent sterol molecule lacks this facility and must therefore be solubilized by interaction with a number of amphipathic agents. In plasma it exists in association with phospholipid and certain proteins to form the lipid–protein complexes that we call lipoproteins. In bile, on the other hand, bile salts and phospholipid act as its emulsifying agents, leading to the production of micellar aggregates. These two transport systems are subject to interdependent regulation in the liver (Shepherd and Packard, 1987). Disturbances of either lead to pathological sequelae which result in deposition of the sterol in unusual sites like the gallbladder and the intima of artery walls. Such deposits are highly resistant to mobilization and further accretion leads to the macroscopic lesions which we recognize as gallstones and atherosclerotic plaques.

Examination of the atherosclerotic plaque reveals the abundant presence of cholesterol in various forms: crystals of cholesterol monohydrate, extracellular perifibrillar lipid and intracellular cholesteryl ester droplets. Where this lipid comes from, what causes its accumulation and how it can be removed are questions that have received much attention over the last 30 years, and they

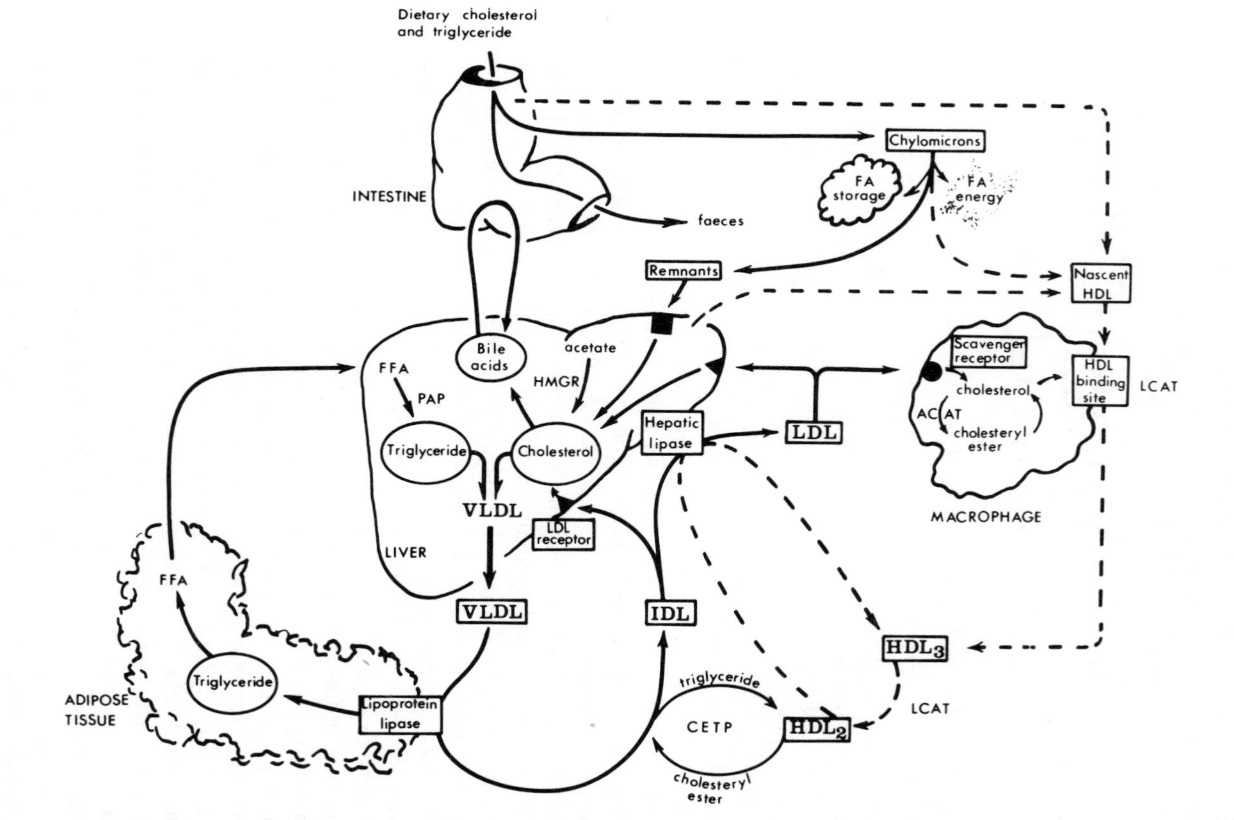

Figure 3.1 Schematic outline of the major elements of human plasma lipoprotein metabolism illustrating the flow of exogenous (dietary) and endogenous cholesterol. Abbreviations: ACAT = acyl CoA-cholesterol acyltransferase; LCAT = lecithin-cholesterol acyltransferase; HMGR = HMG CoA reductase; PAP = phosphatidic acid phosphatase; CETP = cholesteryl ester transfer protein; FA = fatty acids.

continue to be areas of active interest. There is little doubt that the majority of the sterol in artery walls is deposited as a by-product of plasma lipoprotein metabolism. The metabolic fate of lipoproteins is governed by certain key elements: apolipoproteins, cell membrane receptors and plasma enzymes. When these fail to operate efficiently, or when the system is overloaded by excessive dietary fat intake, misdirection of even small amounts of cholesterol can lead, over years, to the genesis of atherosclerosis.

3.2 CHOLESTEROL IN THE PLASMA

Two major plasma protein families, apolipoproteins A and B, are intimately involved in transporting cholesterol (Fig. 3.1) through the plasma (Havel *et al.*, 1980). The former, the main protein of plasma high-density lipoproteins (HDL, density (d) = 1.063–1.21 kg/l), is an avid cholesterol acceptor and is believed to promote the removal of the sterol from cells and mediate its transport centripetally to the liver. Apolipoprotein B (apoB), on the other hand, is an integral component of chylomicrons, very low, intermediate and low-density lipoproteins (VLDL, IDL and LDL). It is secreted with its cholesterol and triglyceride load, delivering these lipids to peripheral parenchymal cells which use them for structural and metabolic purposes.

Chylomicrons are formed during intestinal assimilation of exogenous lipid and are transported in chyle via the systemic circulation to the liver for storage or further processing. Elaboration of these particles is critically dependent on the production of apolipoprotein B which, in the intestine, involves a unique mRNA editing process (Powell *et al.*, 1987). The human genome contains only a single copy of the apoB gene. In the liver the gene is fully expressed (generating $apoB_{100}$). However, intestinal deamidation of the apoB messenger introduces a stop codon (a CAA→ UAA conversion), truncating the apoB to a 2152 amino acid product ($apoB_{48}$) and eliminating the LDL receptor binding site from the protein. The biological benefits inherent in this process are not clear although, arguably, the mechanism is used to alter the metabolism of particles containing B_{48} by inhibiting their ability to bind to the LDL receptor and redirecting their catabolism into alternative degradative pathways.

When the chylomicrons appear in the plasma they acquire apolipoprotein E and the C proteins from HDL and have a density <0.95 kg/l. Apo CII modulates lipolysis of the chylomicron by acting as a cofactor for lipoprotein lipase located on capillary endothelial surfaces in skeletal muscle and adipose tissue (Breckenridge *et al.*, 1978). This enzyme, which is under strong hormonal control, promotes hydrolysis of triglyceride from the particle core, releasing fatty acids for storage or energy requirements. The chylomicron itself shrinks progressively, shedding its surface phospholipid and small

Table 3.1 Plasma lipids and lipoproteins in coronary artery disease (CAD) negative and positive subjects

Group	*LDL cholesterol (mmol/l)*	*HDL cholesterol (mmol/l)*	*HDL_2 mass (mg/dl)*	*HDL_3 mass (mg/dl)*
CAD negative (n = 18)	3.98 ± 0.85	1.18 ± 0.27	46 ± 26	249 ± 42
CAD positive* (n = 24)	23% ↑ $p<0.01$	17% ↓ $p<0.01$	33% ↓ $p<0.05$	14% ↓ $p<0.01$

* ↓ = decrease from CAD negative
↑ = increase from CAD negative

molecular weight proteins into the HDL density interval, to leave a remnant which by virtue of the apoE on its surface is cleared rapidly by the parenchymal cells of the liver. The latter possess an LDL receptor-like polypeptide (LDL receptor-related protein or LRP) which has all the properties expected of the putative chylomicron receptor (Beisiegel *et al.*, 1989). It binds apoE but not apoB, and is insensitive to sterol regulation.

In this regard, the apoE phenotype of individuals has a strong impact on their ability to clear chylomicron remnant particles. The presence of E_2 (results from a mutation in the E protein), even in a single dose, significantly retards chylomicron remnant clearance. Recent studies have suggested that coronary artery disease may be associated with a relative inability to clear chylomicron remnants efficiently from the circulation (Simpson *et al.*, 1989). Administration of a triglyceride-enriched meal to individuals with angiographically proven coronary artery disease (CAD) revealed that these subjects metabolized triglyceride-rich lipoprotein particles with reduced efficiency in comparison to age-matched healthy controls. Interestingly, the fasting plasma lipid and lipoprotein levels (Table 3.1) also showed significant differences from the control subjects. In particular, their circulating LDL cholesterol values were 23% higher and their HDL cholesterol 17% lower than the CAD-negative individuals. The reduction in HDL affected both subfractions. The masses of HDL_2 and HDL_3 in the circulation of the ischaemic subjects were 33% and 14% less than in the controls, respectively. Non-lipid risk factors like blood pressure and Quetelet's index did not show any differences between the groups. These observations raise the interesting possibility that defects in the handling of triglyceride-rich particles might be linked to alterations in the metabolism of LDL and HDL.

The two-step chylomicron clearance pathway delivers triglyceride and cholesterol to the liver where it may be stored or resecreted as components of endogenous lipoproteins. This organ also participates in the continuous elaboration of triglyceride-rich VLDL particles which lie within the density

Table 3.2 Regulation of plasma low-density lipoprotein (LDL) concentration

Factors which raise plasma LDL cholesterol	*Factors which lower plasma LDL cholesterol*
Dietary cholesterol	Carbohydrate-rich diets
Dietary saturated fats	Dietary polyunsaturated fat
Thyroxine deficiency	Vegetarianism
Diabetes mellitus	Insulin
Corticosteroids	Oestrogen
Bile acid administration	Specific lipid-lowering drugs

range 0.95–1.006 kg/l. These contain apolipoproteins B_{100}, C, and E on their surface and are subject to the same hydrolytic process as the chylomicrons. Lipolysis here results through loss of triglyceride in the production of a relatively cholesteryl-ester-rich, intermediate-density, remnant lipoprotein particle (IDL, d = 1.006–1.019 kg/l). In normal individuals about one half of these remnants are cleared directly by the liver while the remainder continue down the delipidation cascade to form LDL (d = 1.019–1.063 kg/l), the major cholesterol transporter in human plasma.

About 70% of the cholesterol which circulates in the bloodstream of normal subjects is associated with LDL whose plasma concentration is positively correlated with CHD risk. Clearly, therefore, basic comprehension of the regulation of LDL metabolism is fundamental to our appreciation of the pathogenesis of atherosclerosis. A number of dietary, humoral and pharmacological perturbations produce major alterations in its plasma level (Table 3.2). Additionally, several genetic derangements of lipoprotein metabolism profoundly affect circulating LDL and have led to a deepening in our understanding of the regulatory factors which govern the rates of synthesis and catabolism of the lipoprotein. Familial hypercholesterolaemia (FH) is an excellent example. This autosomal codominant condition is characterized by a significantly raised plasma LDL level, by xanthomatous accretions of cholesterol at sites predisposed to trauma (such as elbows, knees, tendons) and by premature and accelerated vascular disease (Goldstein and Brown, 1989). The basic lesion in these patients has been localized in the gene coding for a high molecular weight membrane receptor protein. Normally this protein binds LDL at specialized sites on the cell surface and promotes its internalization and subsequent catabolism in secondary lysosomes (Fig. 3.2). The cholesterol liberated into the cytoplasm by this process replenishes the intracellular sterol pool and downregulates both *de novo* endogeneous cholesterol synthesis and further assimilation from the environment (Goldstein and Brown, 1989).

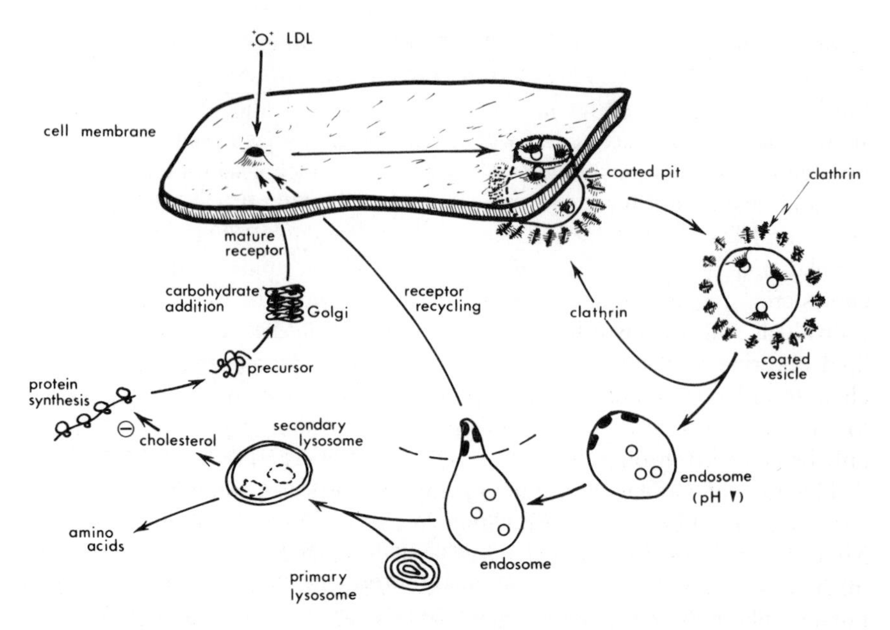

Figure 3.2 The LDL receptor pathway. Following its binding to specific receptors located on cell membranes, LDL is assimilated by endocytosis and directed within endosomes to liposomal degradation. The cholesterol released in the process down regulates the production of new receptors and at the same time suppresses synthesis of HMG CoA reductase, the rate-limiting enzyme for cholesterol production. By this process, the delivery of exogenous sterol inhibits cholesterol synthesis within the cell.

The detailed processes involved here are not known but there is evidence that intracellular cholesterol sufficiency depresses synthesis both of the receptor protein and of 3-hydroxy-3-methylglutaryl coenzyme A (HMG CoA) reductase, the rate-limiting enzyme in cholesterol synthesis. Metabolic studies in normal subjects have indicated that this receptor pathway accounts for about 50% of overall LDL catabolism (Shepherd *et al.*, 1979; Packard *et al.*, 1983). The mechanisms responsible for the rest have not yet been determined but are believed to hold the key to the role of LDL in atherogenesis since, unlike the receptor pathway, their operation is not tightly governed by sterol accumulation in tissues (Brown and Goldstein, 1983). Hence, they continue to operate even when cholesterol has risen in the cells to toxic levels.

3.3 CHOLESTEROL AND ATHEROSCLEROSIS

Recent national surveys of coronary risk factors have shown that about 70% of middle-aged men and women have cholesterol levels above the recommended value (Shaper *et al.*, 1985; Study Group, European Atherosclerosis Society, 1987; Mann *et al.*, 1988) of 5.2 mmol/l, while one-third have a concentration in excess of 6.5 mmol/l and are considered at significant risk of ischaemic heart disease. These findings have prompted calls for a general population-based strategy for lowering cholesterol levels through diet and health education. However, for some of those at the top end of the cholesterol distribution such an approach is inadequate. One in 500 of the population suffers from FH and has a two- to three-fold elevation of plasma cholesterol and severe, premature cardiovascular disease (Slack, 1969). Individuals with the disorder do not respond well to diet, and usually require lipid-lowering drugs to reduce their CHD risk significantly. Since high cholesterol levels do not generally result in signs or symptoms of disease prior to the appearance of coronary problems, the detection of subjects at risk can only be achieved through large-scale screening exercises.

The issue of coronary screening for raised cholesterol and other risk factors (smoking, raised blood pressure, family history, etc.) is at present contentious. Models (Anggard *et al.*, 1986) have been developed in which much of the initial risk factor assessment and follow-up is handled by specially trained nurses. These systems, however, are expensive to operate and need to be evaluated carefully before they are put into widespread use. Cholesterol measurement by itself has only limited ability to identify those at risk, the difference in CHD incidence between the top and bottom quintiles of the cholesterol distribution being only 2–2.5-fold (Study Group, European Atherosclerosis Society, 1987; Mann *et al.*, 1988). Better discrimination is obtained if the lipoprotein profile of VLDL, LDL and HDL is assessed. Each fraction carries cholesterol but has a different impact on the atherogenic process. Furthermore, techniques have been developed recently to subfractionate each of the main plasma lipoprotein classes on the basis of size and density (Griffin *et al.*, 1990). Application of these methods has revealed an unexpected complexity in lipoprotein structure and function, raising the possibility that the atherogenic or antiatherogenic properties of a fraction may reside in one of these specific subpopulations of particles (Griffin *et al.*, 1990).

The discussion which follows describes advances in our understanding of the role that lipoproteins and their subfractions play in atherogenesis. Increasing knowledge in this area will improve our ability to identify the patient at risk for CHD, and to tailor treatment for his or her particular problem.

3.4 LOW-DENSITY LIPOPROTEIN AND ATHEROSCLEROSIS

The level of LDL, the major cholesterol carrier in plasma, is positively correlated with the incidence of CHD. Furthermore, clinical trials have shown that when the concentration of this lipoprotein is reduced there is a lack of progression or even a regression of atherosclerosis (Blankenhorn *et al.*, 1987), and fewer signs and symptoms of CHD. As noted above, apolipoprotein B on the LDL particle's surface facilitates the transport process by interacting with specific high-affinity receptors on cell membranes. These promote endocytosis of the lipoprotein, leading to hydrolysis of its component parts and release of cholesterol for the cell's needs. When sufficient sterol is present in a cell, synthesis of the LDL receptor is suppressed (Goldstein and Brown, 1989).

Transport of cholesterol by this mechanism operates throughout the animal kingdom and in several human populations (such as the Japanese) (Ueshima *et al.*, 1982) at relatively low plasma LDL cholesterol concentrations, i.e. 0.5–2.0 mmol/l. If it is overburdened with sterol through excessive dietary intake, or if a genetic disease such as FH disrupts its efficient functioning, the receptor pathway is shut down and the lipoprotein is redirected to receptor-independent, potentially atherogenic, catabolic routes. The nature of the latter is unknown, although there is evidence to suggest that cells of the monocyte-macrophage series are active in the degradation of LDL by non-receptor mechanisms (Slater *et al.*, 1982; Brown and Goldstein, 1983). These cells are, of course, also believed to play an important role in the early stages of atherogenesis (Fagiotto *et al.*, 1984) and so their interaction with lipoproteins has been intensively studied both *in vitro* and *in vivo*.

Initial experiments showed that macrophages exposed to high concentrations of plasma LDL did not avidly assimilate the lipoprotein or store cholesterol (Goldstein *et al.*, 1979), and so it was suggested that LDL had to be altered in some way before it was recognized by these cells. Further studies found that a modification of potential physiological significance was induced by endothelial cells; LDL incubated with endothelial monolayers for 24–48 hours undergoes oxidation, and this renders it susceptible to uptake by macrophages (Henricksen *et al.*, 1983). These *in vitro* observations suggested a sequence of events that might lead to cholesterol deposition in the artery wall, that is, LDL passing through endothelial cells in the process of transcytosis would be oxidized and hence avidly assimilated by macrophages present in the intima. If the plasma LDL concentration was high enough, this pathway might proceed at a rate sufficient to generate cholesterol-laden foam cells in the subendothelial layer. These would form the focus of a growing lesion (Ross, 1981).

This picture of the role of LDL in the atherosclerotic process has been complicated by the demonstration that the lipoprotein class actually comprises a small number of overlapping species that vary in size and density. At least three subfractions (LDL-I to LDL-III) can be isolated from the plasma of normal and hyperlipidaemic subjects (Beltz *et al.*, 1987). The largest and least dense, LDL-I, is present in high concentration in low-risk individuals like young women, while LDL-III (the smallest, densest particles) are abundant in patients suffering from CHD (Griffin *et al.*, 1990). Metabolic studies indicate that larger LDLs are catabolized more readily by receptors, while a disproportionate amount of smaller LDL is channelled to a receptor-independent mechanism. The latter, therefore, appear to represent a particularly atherogenic variety of LDL. Including their measurement in risk assessment may help explain why subjects with apparently similar general risk profiles and total LDL cholesterol levels can develop CHD at widely differing rates.

It has become clear from recent studies (Eisenberg *et al.*, 1984), that hypolipidaemic drugs, as well as reducing plasma cholesterol and triglyceride levels, have an impact on the structure and function of LDL particles. Bezafibrate therapy in hypertriglyceridaemics, for example, results in an increase in the average size of LDL and a change in the conformation of apolipoprotein B so that it exhibits a higher affinity for receptors. An explanation for these effects is that the drug increases the proportion of LDL-I to LDL-III in the patient's plasma. According to the above argument this should reduce the atherogenicity of the lipoprotein fraction.

Lipoprotein(a) (Lp(a)) is an LDL-like particle isolated in the density range between LDL and HDL (i.e. 1.05–1.08 kg/l). It is emerging as a new, independent risk factor for CHD, and certain aspects of its structure suggest how it may act as such. The unique feature of Lp(a) is the apolipoprotein(a) moiety which is attached to an LDL particle via a disulphide bond. This protein is of high molecular weight and has a remarkable homology with plasminogen. It contains about 37 sulphydryl-linked 'kringle' structures and an inactive 'protease' region (McLean *et al.*, 1987). The similarity to an important component of the thrombolytic process raises the possibility that Lp(a) may increase atherosclerosis risk not only because of its LDL-like properties, but also by interfering with the activation of plasminogen. Evidence for such an effect has been published (Miles *et al.*, 1989). The importance of Lp(a) as an independent risk marker for CHD was highlighted in a study in FH patients (Houlston *et al.*, 1988). It was found that the presence of raised Lp(a) concentrations discriminated between those who had and those who did not have clinical signs and symptoms of coronary disease. We do not know what regulates the concentration of the lipoprotein in plasma, but it appears to be under strong genetic control (Utermann *et al.*, 1987).

3.5 HIGH-DENSITY LIPOPROTEIN AND ATHEROSCLEROSIS

High-density lipoprotein functions as the initiator of 'reverse cholesterol transport', promoting the clearance of sterol from peripheral tissues to the liver (Fig. 3.3). This role underlies the inverse relationship between the plasma level of the lipoprotein and CHD. Many subjects with CHD have low

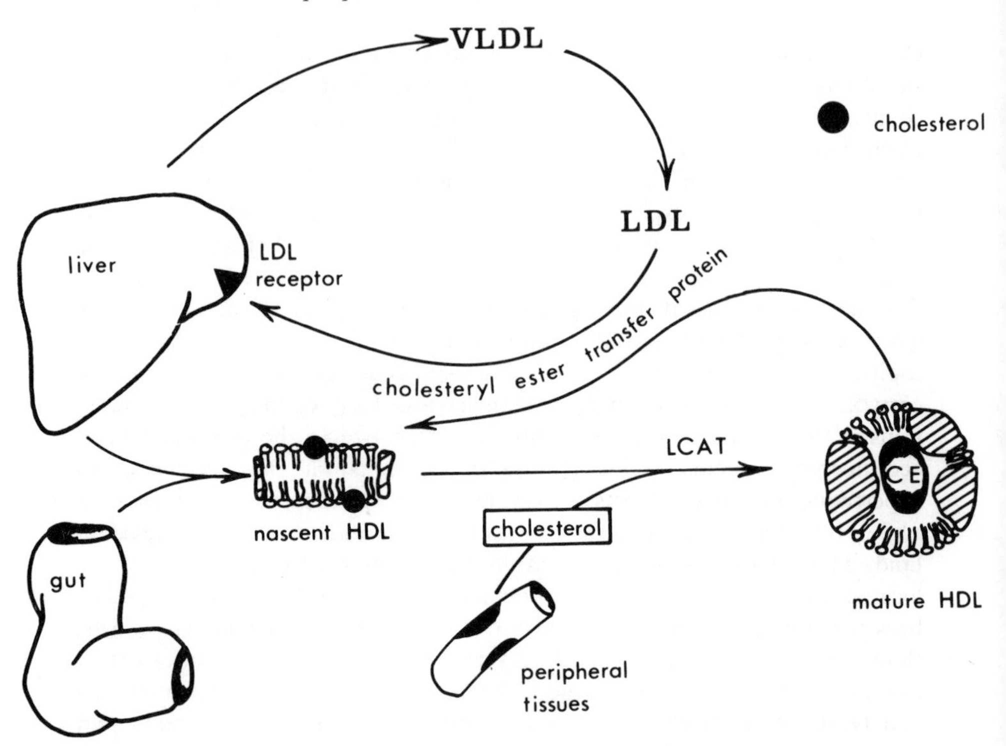

Figure 3.3 The role of HDL in reverse cholesterol transport. Lecithin-cholesterol acyltransferase (LCAT), circulating in association with HDL, traps tissue cholesterol within the particle by esterifying the sterol, increasing its hydrophobicity and promoting its migration to the core of the lipoprotein. The mature HDL generated in the process is then capable of transferring its cholesteryl ester load into chylomicrons and VLDL via the agency of cholesteryl ester transfer protein. These particles, by delivering the cholesteryl ester to the liver, may therefore represent the final pathway for sterol excretion into bile.

HDL cholesterol, often as the only lipoprotein abnormality present. However, it usually appears in association with a raised plasma triglyceride concentration. In fact, the metabolism of HDL is closely linked to that of the triglyceride-rich lipoproteins; they exchange both 'core' and 'coat' components and, to a large extent, HDL is derived from the lipolysis of chylomicrons and VLDL. Cholesterol is mobilized from tissues when HDL binds to the cell membrane and incorporates excess unesterified sterol into its surface monolayer. The nature of the binding site is controversial, and although an HDL-binding polypeptide has been isolated (Barbaras *et al.*, 1990) it is uncertain whether it is a specific protein receptor as in the case of LDL (Karlin *et al.*, 1987).

In vivo, free cholesterol acquired by HDL is rapidly esterified by lecithin-cholesterol acyl transferase (LCAT). The action of this enzyme traps the sterol moiety in the core of the lipoprotein and permits the capture of more cellular cholesterol. Repeated cycles of absorption and esterification lead to expansion of the HDL core and an increase in particle size. As a result, smaller HDLs in the HDL_3 subfraction density range are converted to the larger, less dense HDL_2.

The fate of cholesterol accumulated in the HDL fraction is not entirely clear. In some species, like the rat, HDL is thought to be cleared directly by the liver; however in others, including man, another plasma factor, cholesteryl ester transfer protein (CETP), operates to transfer ester from HDL to triglyceride-rich lipoproteins (Fig. 3.3). These may be the final vehicles that carry cholesterol back to the liver (Albers *et al.*, 1984). The extent of the transfer is dependent on both the activity of CETP and the concentration of acceptor lipoproteins (Fielding and Fielding, 1981). For example, in hypertriglyceridaemic subjects, VLDL has a high content of cholesteryl ester while HDL receives triglyceride in exchange, and its core is enriched in this lipid. This renders HDL susceptible to lipolysis by hepatic lipase, an enzyme located on hepatic sinusoidal endothelial cells. It is able to convert HDL_2 back to HDL_3 and is a major regulator of the HDL_2 concentration in plasma (Kuusi *et al.*, 1980). The higher level of HDL_2 in women is to a large extent attributable to their lower hepatic lipase activity. The activity of the enzyme and HDL cholesterol levels are influenced by sex hormones and by certain hypolipidaemic drugs (Kuusi *et al.*, 1980).

Both low and high HDL levels can be inherited. Recent studies from this and another laboratory have uncovered a strong genetic link between variation in the CETP gene and HDL cholesterol. We examined DNA from subjects with high (>2.0 mmol/l) and low (<1.0 mmol/l) HDL levels. Restriction fragment length polymorphisms (RFLPs) were observed in the genes coding for apolipoprotein-AI, LCAT and CETP. None of the AI or LCAT variants were related to HDL cholesterol, but the TaqIB RFLP in the CETP gene was strongly linked to both the cholesterol and AI (Kondo *et al.*, 1989) components of HDL. Subjects with the cutting site for the enzyme were likely to have higher plasma concentrations of HDL cholesterol and apo AI. Interestingly, this inherited tendency to high HDL levels is abolished by cigarette smoking (Kondo *et al.*, 1989).

3.6 DIET AND PLASMA CHOLESTEROL METABOLISM

Most subjects who die of heart disease do not have a genetically defined dyslipidaemia. In fact, elevation of plasma LDL is more likely to derive from environmental (i.e. dietary) influences. These, interacting with poorly defined, possibly polygenic factors, lead to moderate accumulation of LDL in the circulation to levels significantly in excess of physiological requirements (Goldstein and Brown, 1977). The effects of diet on plasma lipids have been investigated extensively over recent years.

3.6.1 Dietary cholesterol

There is a clear relationship between the amount of cholesterol in the diet and the level of that lipid in the circulation. In a study performed in Glasgow the diets of seven healthy volunteers were supplemented with 1.5 g egg yolk cholesterol/day over a period of 1 month (Packard *et al.*, 1983). This resulted in a 40% increase in circulating LDL. The composition and size of the lipoprotein particles did not change, indicating that their number had risen in the plasma. Kinetic analysis showed that this resulted from a 23% increase in LDL production. The rate of catabolism fell, but only by a small amount (12%) and this was attributable to a reduction in LDL receptor activity. The mechanism behind these changes can be viewed as the response of the liver to expanded delivery of the sterol in chylomicrons.

The gut absorbs a constant fraction of the cholesterol presented to it each day and therefore would transfer an increased load to the liver during egg feeding. This organ, in theory, could compensate by several mechanisms. First, hepatic uptake of lymph chylomicrons could be suppressed. Available evidence, however, indicates that the chylomicron receptor on liver cell membranes is not down regulated by overloading the system with lymph cholesterol (Mahley *et al.*, 1981). In fact, the liver seems to be able to remove any concentration of chylomicron remnants from the plasma. Rather, when faced with this problem of excessive delivery the organ adopts two other mechanisms to deal with the oversupply: it suppresses endogenous cholesterol synthesis (Dietschy, 1984) and at the same time increases its rate of export of the sterol into the systemic and enterohepatic circulations.

This kind of response certainly would explain the increased secretion rate of the lipoprotein into the plasma in our study (Packard *et al.*, 1983); in addition, Quintao *et al.* (1971) have shown that during cholesterol feeding, neutral steroid output into the bile rises significantly. Of course, LDL is not secreted as such into the plasma from the liver but appears initially as its less dense triglyceride-enriched precursor VLDL. Nestel and co-workers (1982) in fact demonstrated a rise in the production rate of this particle when they fed dietary cholesterol supplements to patients. One final point to note from the egg feeding study is that the treatment did result in some down regulation of corporeal (presumably hepatic) LDL receptors, but the extent of this response was smaller than would have been expected had it been the prime mechanism for suppressing hepatic sterol accumulation.

So, overall, dietary cholesterol loading leads to a change in LDL synthesis rather than in catabolism, emphasizing the role of synthetic mechanisms in the control of the plasma LDL concentration. This concept accords with the statistical relationships which have been observed between the plasma LDL pool size and its rates of secretion into, and catabolism from, the plasma (Packard and Shepherd, 1983). It seems that these two mechanisms are regulated independently in man. For example, both neomycin and the bile acid sequestrant resin cholestyramine lower plasma LDL cholesterol by about the same amount (25%), but kinetic studies have shown that this is achieved in different ways. Cholestyramine, as shall be outlined later, activates LDL catabolism without affecting synthesis while neomycin has the opposite effect, suppressing synthesis but leaving catabolism unchanged (Kesaniemi and Grundy, 1984). By blocking cholesterol absorption the antibiotic would be expected to deplete the 'hepatic sterol pool' which, from the foregoing argument, should reduce LDL synthesis. Cholestyramine also reduces the hepatic sterol pool but in this instance the liver responds by increasing its receptor-mediated assimilation of the lipoprotein. The paradox may be explained on the basis of compartmentalization of cholesterol in the liver. Indeed, such an hypothesis was mooted years ago (Balasubramaniam *et al.*, 1973) to account for the observation that when labelled cholesterol was given to an animal the specific activities of biliary neutral sterol, bile acids, hepatic cholesteryl esters and plasma cholesterol were all different. One pool, it may be envisaged, feeds lipoprotein synthesis while the other is responsible for regulation of the receptor.

3.6.2 Dietary fat saturation level and lipoprotein metabolism

Although dietary sterol has an impact on plasma cholesterol, the effect of ingested triglyceride on this parameter is quantitatively more important. The average daily intake of neutral fat is about 120 g, of which only 0.5 g is cholesterol. The dietary triglyceride saturation level perturbs the LDL cholesterol concentration, which rises significantly when saturated animal fat is ingested in substantial amounts. VLDL and HDL and also affected, but to a lesser extent. Numerous studies have shown that both the quantity and saturation level of triglyceride in the diet are potent modulators of lipoprotein metabolism. The mechanisms involved in all of these processes are complex, although it appears that they share a common final pathway. This would certainly explain why the hypercholesterolaemic influence of dietary sterol is blunted by the simultaneous administration of polyunsaturated vegetable or fish oils. Dietary saturated fat, on the other hand, exaggerates the effect of cholesterol feeding (Schonfeld *et al.*, 1982).

We have examined the influence of dietary fat saturation level on plasma lipoprotein metabolism (Shepherd *et al.*, 1980). Subjects were fed isocaloric diets containing 40% of the total calories as fat. During the first phase of the study the polyunsaturated/saturated fat ratio of the diet was maintained at 0.25, while in the second phase it was raised to 4.0. Such a change lowered plasma cholesterol by 23%. The greatest fall was seen in LDL, although substantial reductions in both VLDL and HDL were also recorded. Examination of the composition of the LDL and of the kinetic parameters of its turnover showed that (1) the hypocholesterolaemic response varied from patient to patient and (2) multiple mechanisms were apparently involved. The fall in LDL cholesterol came in part from a reduction in the percentage of sterol in the particle and partly from an increase in the rate of clearance of the lipoprotein from the plasma. This is in distinct contrast to the influence of dietary sterol on LDL turnover (Packard *et al.*, 1983). We could find no obvious change in faecal neutral and acidic steroid excretion in these subjects, nor in their cholesterol balance, and so we were unable to ascribe the changes we observed to a specific effect, say, on the LDL receptor.

Polyunsaturated fat feeding also lowered HDL (Shepherd *et al.*, 1978), reducing the level of its major protein apo AI by 21%. This was attributable to depression of its synthesis. The chemical composition, thermotropic properties and subfraction distribution of the lipoprotein were also affected. The particles become denser, more fluid and relatively depleted in HDL_2. There was no obvious explanation for these findings at the time, although since then Rudel and Parks (1982) have suggested that chylomicrons produced in polyunsaturated-fat fed animals are larger in size and so can transport much more core material per unit of surface available. This surface material, which contains apo AI, is precursor to HDL. So, polyunsaturated fat feeding may perturb HDL metabolism indirectly via its influence on the genesis of chylomicrons in the intestinal mucosa.

Another means of lowering plasma cholesterol is to replace saturated fat in the diet with carbohydrate (Cortese *et al.*, 1983). This lowers LDL by reducing its rate of synthesis. HDL also falls during this manoeuvre (Blum *et al.*, 1977) and so the effect of the diet on the HDL/LDL cholesterol ratio is minimized. In these circumstances one must therefore balance the benefits of LDL reduction against the potential risks of lowering antiatherogenic HDL. But in practice, communities which eat diets low in total fat and high in carbohydrate or vegetable/fish oils experience less atherosclerosis, supporting the concept that dietary intervention of this kind confers practical if not theoretical biochemical benefit.

3.7 DRUGS AND PLASMA CHOLESTEROL METABOLISM

Pharmacologic intervention will not cure a genetic disease whose treatment at the present time is generally symptomatic. However, FH is different from most conditions in this category since patients with the condition do respond to certain drugs by stimulating any residual receptor activity which they possess. Normally, LDL receptors are substantially down regulated *in vivo*, as evidenced by the fact that it is possible to increase the expression of LDL receptors on the membranes of freshly isolated mononuclear cells by exposing them to lipoprotein-free medium (Goldstein and Brown, 1977; Brown and Goldstein, 1983). The same is true for monocytes from heterozygous FH individuals. So there appears to be significant reserve capacity in these cells for cholesterol assimilation. Presumably this process follows depletion of an intracellular cholesterol pool, leading to release of the inhibition of receptor-mediated LDL catabolism.

Certain hypolipoproteinaemic agents available for clinical use seem to operate by triggering this series of events in the liver. The latter plays a predominant role in cholesterol metabolism for two important reasons. First, it is the only site for sterol excretion from the body and second it possesses more than half the total body complement of LDL receptors. So, any event leading to a perturbation of hepatic cholesterol metabolism would be expected to have a profound effect on plasma LDL cholesterol levels.

Hepatocyte sterol pools can be depleted either by inhibiting cholesterologenesis or by promoting the conversion of cholesterol to its oxidation products, the bile acids. Cholestyramine, a bile acid sequestering agent, binds these acids in the lumen of the gut, prevents their return to the liver in the enterohepatic circulation and releases cholesterol 7α hydroxylase from its normal down-regulated state. Cholesterol hydroxylation is thereby promoted, resulting in depletion of a regulatory sterol pool (Packard and Shepherd, 1982). The cells of the liver respond in two ways. Endogenous cholesterol production is accelerated and, secondly, assimilation of sterol from the plasma via the LDL receptor is promoted.

Unfortunately, the increase in hepatocyte cholesterol synthesis blunts LDL receptor activation and partly dissipates the hypocholesterolaemic action of resin therapy. Addition to the regimen of a second agent, an HMG CoA reductase inhibitor, augments the cholesterol-lowering effect of sequestrant resins (Bilheimer *et al.*, 1983) by blocking the hepatocyte escape mechanism. Combined drug therapy is thus warranted in this circumstance and can be used to control the high LDL cholesterol levels seen in the worst affected heterozygotes. Unfortunately, even this aggressive approach fails to elicit an effective response in receptor-negative FH homozygotes, confirming that the receptor is essential to the lipid-lowering activity of the drugs. The most efficient treatment for such individuals is to transplant receptors into them in the form of a new liver. In a case in which this was done (Starzl *et al.*, 1984) the plasma cholesterol level fell rapidly by 70–80% after recovery from the surgical procedure. This encouraging result demonstrates unequivocally the pivotal role of the liver in the control of plasma cholesterol metabolism.

ACKNOWLEDGEMENTS

This work was undertaken during the tenure of grants from the British Heart Foundation (BHF 87/6 and BHF 87/101) and the Scottish Home and Health Department (SHERT 908). Claire McKerron provided excellent secretarial assistance.

REFERENCES

Albers, J.J., Tollefson, J.H., Chen, C.H. and Steinmetz, A. (1984) Isolation and characterisation of human plasma lipid transfer proteins. *Arteriosclerosis*, **4**, 49–58.

Anggard, E., Land, J.M., Lenihan, C.J., *et al.* (1986) Prevention of cardiovascular disease in general practice: a proposed model. *Br. Med. J.*, **293**, 177–80.

Balasubramaniam, S., Mitropolous, K.A. and Myant, N.B. (1973) Evidence for the compartmentalisation of cholesterol in rat liver microsomes. *Eur. J. Biochem.*, **34**, 77–83.

Barbaras, R., Puchois, P., Fruchart, J.C., *et al.* (1990) Purification of an apolipoprotein A binding protein from mouse adipose cells. *Biochem. J.*, **269**, 767–73.

Beisiegel, V., Weber, W., Ihrke, G., *et al.* (1989) The LDL receptor related protein, LRP, is an apolipoprotein E binding protein, *Nature*, **341**, 162–7.

Beltz, W.F., Young, S.G. and Witztum, J.L. (1987) Heterogeneity in low density lipoprotein metabolism, in *Proceedings of the Workshop on Lipoprotein Heterogeneity* (ed. K. Lippel), NIH Publications No. 87–2646, Bethesda, USA, pp. 215–36.

Bilheimer, D.W., Grundy, S.M., Brown, M.S. and Goldstein, J.L. (1983) Mevinolin and colestipol stimulate receptor mediated clearance of low density lipoprotein from plasma in familial hypercholesterolemia heterozygotes. *Proc. Natl Acad. Sci. USA*, **80**, 4124–8.

Blankenhorn, D.H., Nessini, S.A., Johnson, R.L., *et al.* (1987) Beneficial effects of combined colestipol-niacin therapy on coronary atherosclerosis and coronary venous bypass grafts. *JAMA*, **257**, 3233–40.

Blum, C.B., Levy, R.I., Eisenberg, S., *et al.* (1977) High density lipoprotein metabolism in man. *J. Clin. Invest.*, **60**, 795–807.

Breckenridge, W.C., Little, J.A., Steiner, G., *et al.* (1978) Hypertriglyceridemia associated with deficiency of apolipoprotein CII. *N. Engl, J. Med.*, **298**, 1265–70.

Brown, M.S. and Goldstein, J.L. (1983) Lipoprotein metabolism in the macrophage. Implications for cholesterol deposition in atherosclerosis. *Ann. Rev. Biochem.*, **52**, 223–61.

Cortese, C., Levy, Y., Janus, E.D., *et al.* (1983) Modes of action of lipid lowering diets in man: studies of apolipoprotein B kinetics in relation to fat consumption and dietary fat composition. *Eur. J. Clin. Invest.*, **13**, 79–85.

Dietschy, J.M. (1984) Regulation of cholesterol metabolism in man and in other species. *Klin. Wochenschr.*, **62**, 338–45.

Eisenberg, S., Gavish, D., Oschry, Y., *et al.* (1984) Abnormalities in very low, low and high density lipoproteins in hypertriglyceridaemia. Reversal toward normal with bezafibrate treatment. *J. Clin. Invest.*, **74**, 470–82.

Fagiotto, A., Ross, R. and Harker, L. (1984) Studies of hypercholesterolemia in the non-human primate. I. Changes that lead to fatty streak formation. *Arteriosclerosis*, **4**, 323–40.

Fielding, C.J. and Fielding, P.E. (1981) Regulation of human plasma lecithin: cholesteryl acyl transferase activity by lipoprotein acceptor cholesteryl ester content. *J. Biol. Chem.*, **256**, 2102–4.

Goldstein, J.L. and Brown, M.S. (1977) The low density lipoprotein pathway and its relation to atherosclerosis. *Ann. Rev. Biochem.*, **46**, 897–930.

Goldstein, J.L. and Brown, M.S. (1989) Familial hypercholesterolemia, in *The Metabolic Basis of Inherited Disease* (eds C.R. Scriver, A.L. Beaudet, W.S. Sly and D. Valle), McGraw Hill, New York, pp. 1215–50.

Goldstein, J.L., Ho, Y.K., Basu, J.K. and Brown, M.S. (1979) Binding site on macrophages that mediates uptake and degradation of acetylated low density lipoprotein producing massive cholesteryl ester deposition. *Proc. Natl Acad. Sci. USA*, **76**, 333–7.

Griffin, B.A., Caslake, M.J., Yip, B., *et al.* (1990) Rapid isolation of LDL subfractions from plasma by density gradient ultracentrifugation. *Atherosclerosis*, **83**, 59–67.

Havel, R.J., Goldstein, J.L. and Brown, M.S. (1980) Lipoproteins and lipid transport, in *Metabolic Control and Disease* (eds P.R. Bondy and L.E. Rosenberg), W.B. Saunders, Philadelphia, pp. 393–494.

Henricksen, T., Mahoney, E.M. and Steinberg, D. (1983) Enhanced macrophage degradation of biologically modified low density lipoprotein. *Arteriosclerosis*, **3**, 149–59.

Houlston, R., Quiney, J., Mount, J., *et al.* (1988) Lipoprotein(a) and coronary heart disease in familial hypercholesterolaemia. *Lancet*, **ii**, 405.

Karlin, J.B., Johnson, W.J., Benedict, C.R., *et al.* (1987) Cholesterol flux between cells and high density lipoprotein. *J. Biol. Chem.*, **262**, 12557–64.

Kesaniemi, Y.A. and Grundy, S.M. (1984) Turnover of low density lipoproteins during inhibition of cholesterol absorption by neomycin. *Atherosclerosis*, **4**, 41–8.

Kondo, I., Berg, K., Drayna, D. and Lawn, R. (1989) DNA polymorphism at the locus for human cholesteryl ester transfer protein (CETP) is associated with high density lipoprotein cholesterol and apolipoprotein levels. *Clin. Genet*, **35**, 49–56.

Kuusi, T., Saarinen, P. and Nikkila, E.A. (1980) Evidence for the role of hepatic endothelial lipase in the metabolism of plasma high density lipoprotein_2 in man. *Atherosclerosis*, **36**, 589–93.

McLean, J.W., Tombinson, J.E. and Kuang, W.J. (1987) cDNA sequence of human apolipoprotein(a) is homologous to plasminogen. *Nature*, **300**, 132–7.

Mahley, R.W., Hui, D.Y., Innerarity, T.L. and Weisgraber, K.H. (1981) Two independent lipoprotein receptors on hepatic membranes of dog, swine and man. *J. Clin. Invest.*, **68**, 1197–206.

Mann, J.L., Lewis, B., Shepherd, J., *et al.* (1988) Blood lipid concentrations and other risk factors: distribution, prevalence and detection. *Br. Med. J.*, **296**, 1702–6.

Miles, L.A., Fless, G.M., Levin, E.G., *et al.* (1989) A potential basis for the thrombotic risks associated with lipoprotein(a). *Nature*, **338**, 301–3.

Nestel, P., Tada, N., Billington, T., *et al.* (1982) Changes in very low density lipoproteins with cholesterol loading in man. *Metabolism*, **31**, 398–405.

Packard, C.J., McKinney, L., Carr, K. and Shepherd, J. (1983) Cholesterol feeding increases low density lipoprotein synthesis. *J. Clin. Invest.*, **72**, 45–51.

Packard, C.J. and Shepherd, J. (1982) The hepatobiliary axis and lipoprotein metabolism. *J. Lipid Res.*, **23**, 1081–98.

Packard, C.J. and Shepherd, J. (1983) Low density lipoprotein receptor pathway in man: its role in regulating plasma low density lipoprotein levels. *Atheroscler. Revs*, **11**, 29–64.

Powell, L.M., Wallis, S.C., Pease, R.J., *et al.* (1987) A novel form of tissue specific

RNA processing produces apolipoprotein B48 in intestine. *Cell*, **50**, 831–8.

Quintao, E., Grundy, S.M. and Ahrens, E.H. (1971) Effects of dietary cholesterol on the regulation of total body cholesterol in man. *J. Lipid Res.*, **12**, 233–47.

Ross, R. (1981) Atherosclerosis: a problem of the biology of arterial wall cells and their interactions with blood components. *Arteriosclerosis*, **1**, 293–311.

Rudel, L.L. and Parks, J.S. (1982) Different kinetic fates of apolipoproteins AI and AII from lymph chylomicra of non human primates. Effect of saturated versus polyunsaturated dietary fat. *J. Lipid Res.*, **23**, 410–21.

Schonfeld, G., Patsch, W., Rudel, L.L., *et al.* (1982) Effects of dietary cholesterol and fatty acids on plasma lipoproteins. *J. Clin. Invest.*, **62**, 1072–80.

Shaper, A.G., Pocock, S.J., Walker, M., *et al.* (1985) Risk factors for ischaemic heart disease: the prospective phase of the British Regional Heart Study. *J. Epidemiol. Community Health*, **39**, 197–209.

Shepherd, J., Bicker, S., Lorimer, A.R. and Packard, C.J. (1979) Receptor mediated low density lipoprotein catabolism in man. *J. Lipid Res.*, **20**, 999–1006.

Shepherd, J. and Packard, C.J. (1987) Lipid transport through the plasma: the metabolic basis of hyperlipidaemia. *Ballieres Clinical Endocrinology and Metabolism*, **1**, 495–514.

Shepherd, J., Packard, C.J., Grundy, S.M., *et al.* (1980) Effects of saturated and polyunsaturated fat diets on the chemical composition and metabolism of low density lipoproteins in man. *J. Lipid Res.*, **21**, 91–9.

Shepherd, J., Packard, C.J., Patsch, J.R., *et al.* (1978) Effects of dietary polyunsaturated and saturated fat on the properties of high density lipoproteins and the metabolism of apolipoprotein AI. *J. Clin. Invest.*, **61**, 1582–92.

Simpson, H.S., Williamson, C.M., Pringle, S., *et al.* (1989) Hypolipidemic drugs and chylomicron metabolism, in *Intestinal Lipid and Lipoprotein Metabolism* (eds E. Windler and H. Greten), W. Zuchschwerdt Verlag, Munich, 194–201.

Slack, J. (1969) Risks of ischaemic heart disease in familial hypercholesterolaemic states. *Lancet*, **ii**, 1380–2.

Slater, H.R., Packard, C.J. and Shepherd, J. (1982) Receptor-independent catabolism of low density lipoprotein: involvement of the reticuloendothelial system. *J. Biol. Chem.*, **257**, 307–10.

Starzl, T.E., Bahnson, H.T., Hardesty, R.L., *et al.* (1984) Heart-liver transplantation in a patient with familial hypercholesterolaemia. *Lancet*, **i**, 1382–3.

Study Group, European Atherosclerosis Society (1987) Strategies for the prevention of coronary heart disease: a policy statement of the European Atherosclerosis Society. *Eur. Heart J.*, **8**, 77–88.

Ueshima, H., Iida, M. and Shimamoto, T. (1982) Dietary intake and total serum cholesterol level: their relationship to different lifestyles in several Japanese populations. *Circulation*, **66**, 519–26.

Utermann, G., Menzel, H.J., Kraft, H.G., *et al.* (1987) Lp(a) glycoprotein phenotypes. Inheritance and relation to Lp(a) – lipoprotein concentrations in plasma. *J. Clin. Invest.*, **80**, 458–65.

4 Unobserved publications

D. Kritchevsky

4.1 INTRODUCTION

Atherosclerosis is a multifactorial disease. Jean Davignon (1978) has successfully tried to summarize all aspects of this disease in a cartoon which attempts to provide a unified theory of the aetiology of this disease. He divides the field into three areas. The first, called ecological factors, refers to the outside factors which impinge on atherosclerosis. These include all aspects of the diet, smoking, genetics, stress, viruses, etc. All of these factors are easy to identify and to measure. Next there is the circulating blood which contains hormones, lipoproteins and platelets and whose rheologic properties may also exert an effect. Factors in this area are almost as easy to measure as those on the exterior. Finally there is the arterial wall itself whose metabolism is still not fully elucidated and which is really the arena for the atherosclerotic event. So we have a very complex physiological process which we try to understand, primarily, by studying and analysing one peripheral tissue, the blood. We have learned much about cholesterol transport and have put this acquired knowledge to excellent use, but we still must recognize that blood cholesterol values are subjected to many physiological perturbations. Plasma cholesterol values represent a major risk factor for coronary disease and a risk factor represents a statistical, not a medical, diagnosis.

4.2 BLOOD CHOLESTEROL

There are few who dispute the observation that cholesterol levels above 240 mg/dl (6.20 mmol/l) put subjects at elevated risk of coronary disease. Why, then, the lack of unanimity? After all, the same data are available for everyone's examination. Differences in opinion are due to the background and viewpoint one brings to the reading of the available material. There are a few pieces of the puzzle which are still missing. Is the serum or plasma cholesterol level an 'etched-in-stone' value which serves as an incontrovertible indicator of risk? Actually this level may vary widely and even a small variation may, in the patient's mind, spell the difference between safety and risk (Hegsted and

Nicolosi, 1987). There are a number of papers bearing on cholesterol variation but they are not widely known or cited. Another question which requires a more stringent answer is that of the influence of dietary cholesterol on blood cholesterol. Does the relationship warrant the level of concern which, in some quarters, borders on hysteria?

In a review of data from the Seven Countries Study, Hegsted and Ausman (1988) showed that deletion of dietary intake data from Japan and Italy (the two lowest intakes) would render the linear correlation no longer statistically significant. Miller *et al.* (1990) studied men and women whose cholesterol levels were below 200 mg/dl and who were with or without angiographically demonstrable coronary artery disease. The 139 men with disease exhibited an average plasma cholesterol of 175 ± 20 mg/dl (4.52 mmol/l), whereas the 55 free of disease had an average level of 167 ± 25 mg/dl (4.32 mmol/l), a difference of less than 5%. In 37 women with artery disease the cholesterol level was 176 ± 23 mg/dl (4.55 mmol/l), while 57 women free of disease had a level of 167 ± 26 mg/dl (4.32 mmol/l). This difference is 4%. The data from the males were significant at the 5% level, but those from the females were not. This differences in significance may reflect 194 male subjects versus 94 females. However, one must always ask if statistically significant data are also biologically significant.

Will lowering blood cholesterol reduce that portion of risk attributable to elevated serum cholesterol? A number of clinical trials have been conducted with results which are encouraging. Presentation of data as relative effects (that is, percentage difference between two percentages) may make them look more promising than they really are. Unfortunately, death from all causes is usually the same in both the test group and the controls.

There are data in the literature bearing on all of these points which leads to the reason for the title of this exposition. We have all read papers in which data are referenced as 'unpublished observations' – much of what follows can be considered as 'unobserved publications'.

4.3 VARIABILITY

There are a number of factors which may influence cholesterol levels including stress, position of the subject during phlebotomy, time of day and season of the year. Many workers have observed cholesterol fluctuations over time but they do not fit a strict seasonal pattern. The question of cholesterol variation in animals and man has been reviewed by Kritchevsky (1985) and by Durrington (1990). One example stands out: Dr M.E. Groover was a medical officer at The Pentagon, where for a 5-year period he ministered to the health of 177 generals, and over that period he carried out a number of physical examinations on each general. At the end of this time he calculated the mean cholesterol value (over the 5 years) for each subject and then plotted the extent of variation from the mean versus number of probands. In a majority of the subjects cholesterol levels fluctuated between 20% and 40%. In some the variation was small and in others (37 of 177) the variation was over 50%. In the 5 years of the study there were 16 coronaries, all of which occurred in men whose cholesterol levels had fluctuated by 50% or more (Groover *et al.*, 1960). If we accept the fluctuations as a genuine physiological effect then single determinations of cholesterol may provide no accurate measure of risk. The cause of the fluctuations has not been defined, although stress had been implicated in some instances.

Durrington (1990) cites a graph of seasonal variation of serum cholesterol in the WHO clofibrate trial (taken from a master's thesis in statistics presented to the London School of Economics) which shows peaks in April and November and valleys in February and June of the same year. The change in cholesterol level between the lowest and highest points is 16 mg/dl (0.41 mmol/l).

4.4 DIET

There is an extensive literature bearing on the effects of diet on serum cholesterol and some of it deals with effects of dietary cholesterol. Studies conducted in free-living subjects (which most of us are) show little effect of cholesterol, usually fed as egg yolk.

In 1970, Kannel and Gordon reviewed the data accumulated in the Framingham Project which began in the late 1940s. Four hundred and thirty-one men and 442 women were segregated into three groups based on their serum cholesterol level: below 180 mg/dl (4.65 mmol/l), 180–299 mg/dl (4.65–7.73 mmol/l) or over 300 mg/dl (7.75 mmol/l). There were no differences in intake of calories, protein, carbohydrate, fat or cholesterol. In a later study from the same project, Dawber *et al.* (1982) showed that cholesterol levels were similar in men who ingested 10.6 eggs per week or 1.4 eggs per week, and in women who ingested 7.3 or 0.7 eggs per week. The number of eggs represented the highest and lowest tertiles of intake.

Gertler *et al.* (1950) segregated from a heart disease study four groups of ten men: the ones who had the highest or lowest serum cholesterol levels and the ones who ingested the most or least cholesterol. They were matched with an equal number of controls. In every subgroup the coronary patients exhibited significantly higher cholesterol levels than the controls, but in no case was there any relation to the amount of cholesterol ingested. Thus, in the control group, men who took in 7.0 or 1.4 g cholesterol per week had serum cholesterol levels of 213 ± 11 (5.50 ± 0.28 mmol/l) and 222 ± 16 mg/dl (5.74 ± 0.41 mmol/l), respectively. In the corresponding coronary groups the

Table 4.1 Daily nutrient intake and heart disease (CHD) in three trial locations (after Gordon *et al.*, 1981)

Location	*Framingham*		*Puerto Rico*		*Hawaii*	
	No CHD	*CHD*	*No CHD*	*CHD*	*No CHD*	*CHD*
No. of subjects	780	79	1932	286	7008	264
Total calories	2622	2488	2395	2289	2319	2210*
Protein (g)	101	99	86	85	95	95
Carbohydrate (g)	252	248	280	262*	264	249*
Sugar	72	78	52	50	46	45
Starch	117	118	180	167*	165	155*
Other	61	52*	48	45	52	48
Fat (g)	114	111	95	94	87	86
P/S	0.39	0.41	0.45	0.49	0.54	0.57
Cholesterol (mg)	529	534	417	419	555	549
Alcohol (g)	25	12*	12	8	14	8*

* Significantly different from no CHD

men ingested 5.7 or 1.3 g cholesterol per week and had serum cholesterol levels of 288 ± 22 (7.44 ± 0.57 mmol/l) and 271 ± 14 mg/dl (7.00 ± 0.36 mmol/l).

There are three large, ongoing, prospective coronary disease studies in Framingham, Puerto Rico and Hawaii that include more than 10 000 men. Gordon *et al.* (1981) analysed the diets of the men who did or did not have coronary disease in each locale. Between the two groups there were statistically significant differences in total calories, carbohydrate, starch and alcohol. In each location men with or without coronary heart disease ate the same amount of fat and cholesterol (Table 4.1). McNamara (1990) has reviewed data from 68 clinical studies on the effect of dietary cholesterol on plasma cholesterol in a total of 1490 subjects. A mean change of plasma cholesterol of 2.3 ± 0.2 mg/dl (0.059 mmol/l) was observed for an increment of 100 mg/day of cholesterol.

The saturation level of dietary fat does affect cholesterolaemia. McNamara *et al.* (1987) fed normal volunteers diets low or high in cholesterol with fats which were predominantly saturated or unsaturated. The extent of fat saturation was the determining factor with regard to cholesterolaemia. Thus, when the fat was saturated the difference in cholesterol levels between subjects on low or high cholesterol was 2.1%; when the fat was unsaturated the difference was 2.8%. However, going from unsaturated to saturated fat raised cholesterol levels by 11.5% in the group fed the low cholesterol supplement and by 10.7% in subjects receiving the high cholesterol supplement (Table 4.2). Cholesterol absorption on the saturated fat regimen

Table 4.2 Plasma cholesterol levels in subjects* fed high or low levels of cholesterol with saturated or unsaturated fat (after McNamara *et al.*, 1987)

Dietary fat	*P/S*	*Dietary (mg)*	*Plasma*	
			mg/dl	*mmol/l*
Low cholesterol				
Saturated	0.31 ± 0.18	288 ± 64	243 ± 50	6.28 ± 1.29
Unsaturated	1.90 ± 0.90	192 ± 60	218 ± 46	5.63 ± 1.19
High cholesterol				
Saturated	0.27 ± 0.15	863 ± 161	248 ± 51	6.41 ± 1.32
Unsaturated	1.45 ± 0.50	820 ± 102	224 ± 46	5.79 ± 1.19

* 39 subjects in the unsaturated fat group, 36 in the saturated fat group

was 61.5 ± 12.6% and 53.5 ± 12.6% on the low and high cholesterol intakes, respectively. When the diet was rich in polyunsaturated fat, subjects on the low cholesterol regimen absorbed 60.5 ± 12.6% of the cholesterol fed, whereas those on high cholesterol absorbed 57.2 ± 12.5%. They also found that about two-thirds of their subjects compensated for the dietary cholesterol load by reducing synthesis.

Frantz *et al.* (1989) have described a study in which 9057 institutionalized subjects in Minnesota were given diets whose fat had a P/S ratio of either 1.60 or 0.28 (P/S relates to ratio of polyunsaturated to saturated fatty acids of different fats). The diet fed to the control group (P/S = 0.28) contained 39.1 energy % from fat and 446 mg/day of cholesterol and the diet fed to the test group (P/S 1.60) contained 37.8 energy % from fat and provided 166 mg cholesterol daily. Serum cholesterol levels were unchanged (up 2%) in the subjects on the more saturated fat and fell by 13.8% in those on the unsaturated fat. In the combined (male plus female) treatment group there were 269 deaths, of which 141 were due to cardiovascular disease; whereas in the control group there were 248 deaths, 144 due to cardiovascular disease.

4.5 BLOOD CHOLESTEROL REDUCTION AND OUTCOME

Will lowering serum cholesterol levels definitely reduce the risk of a heart attack? There have been a number of trials bearing on this point. Generally, even when coronary deaths are reduced, total mortality is not. In a review, Rose and Shipley (1990) stated 'Thus reductions in blood cholesterol can be expected to extend life and to decrease deaths from CHD, with a compensating increase in deaths from other causes, including cancers'. It is of little consolation to have proved the hypothesis but to have died anyhow.

Table 4.3 Deaths in selected cholesterol-lowering trials (after Rossouw and Rifkind, 1990)

Trial[a]	*Number*[b]	*Number (and %) of deaths*		
		CHD	*Non-CHD*	*Total*
Los Angeles	T 424	57 (13.4)	85 (20.0)	174 (41.0)
VA (D)	C 422	81 (19.2)	71 (16.8)	177 (41.9)
Oslo (D)	T 206	38 (18.4)	3 (1.5)	41 (19.9)
	C 206	52 (25.2)	3 (1.5)	55 (26.7)
LRC-CPPT (P)[c]	T 1906	37 (1.9)	31 (1.6)	68 (3.6)
	C 1900	47 (2.5)	24 (1.3)	71 (3.7)
Helsinki (P)[d]	T 2051	22 (1.1)	23 (1.1)	45 (2.2)
	C 2031	23 (1.1)	19 (0.9)	45 (2.1)
Coronary Drug Project (P)	T 1103[e]	241 (21.8)	29 (2.6)	281 (25.5)
	C 2789	633 (22.7)	54 (1.9)	709 (25.4)
	T 1119[f]	238 (21.3)	30 (2.7)	273 (24.4)
Stockholm (P)[g]	T 279	54 (19.4)*	7 (2.5)	61 (21.9)*
	C 276	75 (27.2)	7 (2.5)	82 (29.7)
WHO (P)[e]	T 5331	68 (1.3)	94 (1.8)*	162 (3.0)*
	C 5296	68 (1.2)	65 (1.2)	127 (2.3)

[a] D = diet, P = drug; [b] T = treatment, C = control; [c] cholestyramine; [d] gemfibrozil; [e] clofibrate; [f] niacin; [g] clofibrate + niacin;
* $p < 0.05$

The celebrated lipid research clinics who took part in the coronary primary prevention trial (LRC-CPPT) (Lipid Research Clinics Program, 1984a,b) were involved in screening almost half a million men to find the 3806 who qualified for the study criteria, namely 35–59 years old with a blood cholesterol over 265 mg/dl (9.43 mmol/l). The test group comprised 1906 men given a hypocholesterolaemic drug and the remaining 1900 men were controls. After 7 years there were 71 deaths in the placebo group and 68 in the test group. Deaths from coronary heart disease numbered 44 in the placebo group and 32 in the controls. Coronary incidents numbered 187 (9.8%) in the control group and 155 (8.1%) in the test group. The actual difference is 1.7%, the relative difference in risk, that is the difference between 9.8 and 8.1, is 19%. Similarly in the Helsinki Heart Study (Frick *et al.*, 1987), another drug study, the actual numbers of coronary deaths were 22 out of 2051 in the test group (1.07%) and 23 out of 2031 (1.13%) in the controls. Reduction in risk was actually 2.9% and 3.9% in the two groups, the difference between 2.9 and 3.9 is 34%. Rossouw and Rifkind (1990) have summarized data relating to cardiovascular and total deaths in a number of cholesterol-lowering trials (Table 4.3).

The Multiple Risk Factor Intervention Trial (MRFIT) involved the screening of 361 662 men to provide two groups: one of 6428 men who were given advice on diet, smoking, hypertension, etc., and one of 6438 men who received 'usual care'. After more than 7 years, serum cholesterol levels in the test group were 7% below the starting level. Heart attacks in the test group numbered 38 and there were 35 in the untreated group; there were 265 deaths in the treated group and 260 in the controls (MRFIT Research Group, 1982). After 10.5 years, or about 3 years after termination of the trial, deaths from cardiovascular disease numbered 266 in the test group and 290 in the controls ($p = 0.16$); total deaths in the two groups were 496 and 537, respectively ($p = 0.10$) (MRFIT Research Group, 1990).

In a review of data obtained from a number of epidemiological studies, Taylor *et al.* (1987) constructed a model to provide quantitative data for an increase in life expectancy if one adhered to a life-long 'prudent' diet. By their calculation, a non-smoking, normotensive 40-year-old male with an average cholesterol level would increase his life expectancy by 10 days.

If serum cholesterol level is plotted against coronary deaths in the MRFIT cohort the result is a smooth parabolic curve starting from 140 mg/dl (3.62 mmol/l) with the slope changing markedly at about 240 mg/dl (6.20 mmol/l). If serum cholesterol level is plotted against all-cause mortality, a J-shaped curve results with the low point being at about 170–180 mg/dl (4.39–4.65 mmol/l) (Martin *et al.*, 1986).

4.6 LOW BLOOD CHOLESTEROL AND OUTCOME

It is not unreasonable to ask if there are dangers inherent in very low cholesterol levels. Cholesterol is not a foreign substance. It is a ubiquitous body chemical with metabolic and structural functions. Can depletion of cholesterol carry an inherent risk? There have been many studies relating to low cholesterol levels as a risk for cancer and there has been no resolution. McMichael *et al.* (1984) thoroughly reviewed the data and concluded that even if one assumed that a number of the subjects had low cholesterol levels as a consequence of their tumour, a problem existed which required attention and explanation. Two other publications merit attention. Cowan *et al.* (1990) examined the association between serum lipids and lipoproteins in 2753 men and 2476 women who participated in the Lipid Research Clinics Program Mortality Follow-up Study. They found that men, but not women, in the lowest quartile of serum cholesterol (187 mg/dl or 4.83 mmol/l or lower) or of low-density-lipoprotein cholesterol (119 mg/dl or 3.07 mmol/l or lower) were at greatly increased risk of colon cancer. The authors emphasize that the lipid levels observed were not due to pre-existing disease. Reed (1990) examined the paradoxical situation of Asian populations with a low risk of coronary heart disease but a high risk of stroke. He concluded that the situation was associated with the following variables: low serum cholesterol levels, high alcohol intake and low intake of fat and protein from animal sources.

4.7 CONCLUSIONS

Elevated serum cholesterol is an important and independent risk factor for coronary heart disease, but the additive nature of the other major risk factors (cigarette smoking, diastolic blood pressure above 90) must not be overlooked. The levels of LDL and HDL and their ratio also play an important role, as do levels of apolipoproteins. But the most important determinant is genetics. Since 1968, mortality from heart disease has declined steeply in the USA and a decline has been noted in other countries as well. The reasons are still unclear. One review (Sytkowski *et al.*, 1990) has suggested that this phenomenon is due to improved survival among new cases rather than greatly decreased incidence. In a study of men within the Framingham Heart Study who were free of disease at baseline in 1950, 1960 or 1970 the differences in serum cholesterol between 1950 and 1970 were 7 mg/dl (228 $\pm$ 40 versus 221 $\pm$ 38) or 0.18 mmol/l; there were 39% fewer smokers and 29% fewer hypertensives. In view of the increased affluence in the USA over that period and of the findings of Barker *et al.* (1989) regarding low birthweight as a risk factor for cardiovascular disease it would be interesting to know more about the very early history of the probands.

In the absence of an unequivocal test for the impending coronary we must rely on the measurement of risk factors for identification of those susceptible to coronary heart disease. It is important to be aware of the weaknesses of these measurements as well as of their strengths.

ACKNOWLEDGEMENT

This work was supported, in part, by a grant (HL03299) and a Research Career Award (HL00734) from the National Institutes of Health (US) and by funds from the Commonwealth of Pennsylvania.

REFERENCES

Barker, D.J.P., Winter, P.D., Osmond, C., *et al.* (1989) Weight in infancy and death from ischaemic heart disease. *Lancet*, **i**, 577–80.

Cowan, L.D., O'Connell, D.L., Criqui, M.H., *et al.* (1990) Cancer mortality and lipid and lipoprotein levels. The Lipid Research Clinics Program Mortality Follow-up Study. *Am. J. Epidemiol.*, **131**, 468–82.

Davignon, J. (1978) The lipid hypothesis: pathophysiological basis. *Arch. Surg.*, **113**, 28–34.

Dawber, T.R., Nickerson, R.J., Brand, F.N. and Pool, J. (1982) Eggs v serum cholesterol, and coronary heart disease. *Am. J. Clin. Nutr.*, **36**, 617–25.

Durrington, P.N. (1990) Biological variation in serum lipid concentrations. *Scand. J. Clin. Lab. Invest.*, **50** (suppl. 198), 86–91.

Frantz, I.D. Jr, Dawson, E.A., Ashman, P.L., *et al.* (1989) Test of effect of lipid lowering by diet on cardiovascular risk. The Minnesota Coronary Survey. *Arteriosclerosis*, **9**, 129–35.

Frick, M.H., Elo, O., Haapa, K. *et al.* (1987) Helsinki Heart Study: Primary-Prevention Trial with Gemfibrozil in Middle-aged Men with Dyslipidemia. Safety of Treatment, Changes in Risk Factors, and Incidence of Coronary Heart Disease. *N.E.J.M.*, **317**, 1237–45.

Gertler, M.M., Garn, S.M. and White, P.D. (1950) Serum cholesterol and coronary artery disease. *Circulation*, **2**, 696–702.

Gordon, T., Kagan, A., Garcia-Palmieri, M., *et al.* (1981) Diet and its relation to coronary heart disease and death in three populations. *Circulation*, **63**, 500–15.

Groover, M.E. Jr, Jernigan, J.A. and Martin, C.D. (1960) Variations in serum lipid concentration and clinical coronary disease. *Am. J. Med. Sci.*, **239**, 133–9.

Hegsted, D.M. and Ausman L.M. (1988) Diet, alcohol and coronary heart disease in men. *J. Nutr.*, **118**, 1184–9.

Hegsted, D.M. and Nicolosi, R.J. (1987) Individual variation in serum cholesterol levels. *Proc. Natl Acad. Sci. USA*, **84**, 6259–61.

Kannel, W.B. and Gordon, T. (1970) *The Framingham Diet Study: Diet and the Regulation of Serum Cholesterol.* USDHEW, The Framingham Study, Section 24, US Government Printing Office, Washington, DC.

Kritchevsky, D. (1985) Variation in serum cholesterol levels, in *Nutrition Update*, Vol. 2 (eds J. Weininger and G.M. Briggs), John Wiley, New York, pp. 91–103.

Lipid Research Clinics Program (1984a) The Lipid Research Clinics Coronary Primary Prevention Trial Results I. Reduction in incidence of coronary heart disease. *JAMA*, **251**, 351–64.

Lipid Research Clinics Program (1984b) The Lipid Research Clinics Coronary Primary Prevention Trial Results II. The relationship of reduction in incidence of coronary heart disease to cholesterol lowering *JAMA*, **251**, 365–74.

McMichael, A.J., Jensen, O.M., Parken, D.M. and Zaridze, D.G. (1984) Dietary and endogenous cholesterol and human cancer. *Epidemiol. Rev.*, **6**, 192–216.

McNamara, D.J. (1990) Relationship between blood and dietary cholesterol, in *Meat and Health. Advances in Meat Research*, Vol. 6 (eds A.M. Pearson and T.R. Dutson), Elsevier Applied Science, London, pp. 63–87.

McNamara, D.J., Kolb, R., Parker, T.S., *et al.* (1987) Heterogenicity of cholesterol homeostasis in man: response to changes in dietary fat quality and cholesterol quantity. *J. Clin. Invest*, **79**, 1729–39.

Martin, M.J., Hulley, S.B., Browner, W.S., *et al.* (1986) Serum cholesterol, blood pressure and mortality: implications from a cohort of 361 662 men. *Lancet*, **ii**, 933–6.

Miller, M., Mead, L.A., Kwiterovitch, P.O. Jr and Pearson, T.A. (1990) Dyslipidemias with desirable plasma total cholesterol levels and angiographically demonstrated coronary artery disease. *Am. J. Cardiol.*, **65**, 1–5.

Multiple Risk Factor Intervention Trial Research Group (1982) Multiple Risk Factor Intervention Trial Risk Factor Changes and Mortality Results. *JAMA*, **248**, 1465–77.

Multiple Risk Factor Intervention Trial Research Group (1990) Mortality rates after 10.5 years for participants in the Multiple Risk Factor Intervention Trial. Findings related to a priori hypotheses of the Trial. *JAMA*, **263**, 1795–1802.

Reed, D.M. (1990) The paradox of high risk of stroke in populations with low risk of coronary heart disease. *Am. J. Epidemiol.*, **131**, 579–88.

Rose, G. and Shipley, M. (1990) Effects of coronary risk reduction on the pattern of mortality. *Lancet*, **i**, 275–7.

Rossouw, J.E. and Rifkind, B.M. (1990) Does lowering serum cholesterol levels lower coronary heart disease risk? *Endocrinol. Metab. Clin. No. America*, **19**, 279–97.

Sytkowski, P.A., Kannel, W.B. and D'Agostino, R.B. (1990) Changes in risk factors and the decline in mortality from cardiovascular disease. The Framingham Heart Study. *N. Engl. J. Med.*, **322**, 1635–41.

Taylor, W.C., Pass, T.M., Shepard, D.S. and Komaroff, A.L. (1987) Cholesterol reduction and life expectancy. A model incorporating multiple risk factors. *Ann. Intern. Med.*, **106**, 605–14.

DISCUSSION

Wright, Guildford Can I ask about alcohol? One of the things that alcohol does is to increase alimentary hyperlipimea and yet we are told by the same token that it decreases the incidence of coronary heart disease if taken in moderation. Can you square that particular circle?

Kritchevsky Moderation, especially in intake of alcohol, is a purely individual matter and may act just as a sort of relaxation. Framingham booklet 24 showed another aspect of lifestyle and risk of coronary disease. Teetotallers are at higher risk than people who drink moderately and people who drink moderately are at lower risk than people who drink a lot. Part of that is because if you really are a significant drinker you develop high blood pressure and a lot of other things because alcohol is another source of calories. So again you get back to the whole idea of moderation, especially with respect to drinking.

Shepherd One of the things I did not tell you about my studies of the postprandial lipimea was that if you measure the basal fasting triglyceride levels in the individuals about to get the diet you will find in fact that there is a statistical difference between the basal values in the individuals who are controls and those who are CHD positive. I think that may be the key to your question. If you give these two groups alcohol, one will show an increase and the other won't, because one pathway is saturated and the other is not.

Kritchevsky If I may say something in relation to that, McNamara has done a study very much like the ones I showed you. He fed people diets high or low in cholesterol or high or low in polyunsaturated fat and he showed, as everybody else does, that it is the type of fat which seems to make the most difference. What was interesting, however, is that in two-thirds of his population he found they could compensate for a cholesterol load by reducing their own synthesis. It may very well be that the one-third that cannot compensate should be the people that should be sought out and vigorously treated.

McCarron, Oregon We see this risk curve that is projected in all meetings about the relationship between serum cholesterol and risk. It is my understanding that if you took the MRFIT data and you subtracted the diabetics and smokers from that curve, that you would end up with two distinctly different curves. One would have an inflection point well below the level we are now talking about for those who are diabetics and smokers, and for the rest of us who are fortunate enough not to have these two risk factors, the level that has been targeted at least by the US government would now be well out of the bounds of reason. People would have levels of serum cholesterol of 260–270 mg % that would not be associated with any substantial increase in cardiovascular risk. These data bases are big enough that we ought to be able to separate out people with additional risk factors and we should only use this graph after subdividing it into different categories of risk.

Because of our interest in cation intake and the fact that certain cations, specifically calcium and potassium, track with fat in the diet, our group at Oregon is now concerned about differential blood pressure effects of the various dietary fats. There is the rarely studied and overlooked possibility that certain types of fats, while they may have some impact upon serum cholesterol, have divergent effects upon other confounding cardiovascular risks such as blood pressure, which is of course important.

Kritchevsky In relation to the MRFIT study, Bob Olsen did an analysis of it and he found that in those men who only had elevated cholesterol, the death rate was exactly the same as it was for the controls. But as the numbers of other risk factors rose then their death rate rose and the ones that had all three were at an astronomical figure. It is a multifactorial disease and you cannot spend all your time worrying about only one factor.

Shepherd I think you highlighted that very well. That is the dilemma, all the factors are linked together in some way, maybe directly, maybe indirectly, and to strip off some factors is interesting as a means of getting a clue as to what individual things are doing, but in real life they all mix together.

Lund, Norwich You talked about the pool of LDL and the effect of adding extra cholesterol. What do you think this LDL pool level is as a minimum? I presume that you have to have some LDL for the normal functioning of the body.

Shepherd When you are born you have roughly 50 mg/dl cholesterol in the LDL fraction in your blood. As an adult you have at least twice that concentration in your blood, a value that is twice as much as you need to saturate the LDL receptors. Now presumably the LDL receptors are responsible for delivering cholesterol for the biological needs of cells to grow, divide and make sex hormones and because of that, the value that you would need to saturate these levels is half of the adult.

Kritchevsky It should also be pointed out, and some of Dr Shepherd's research shows this, that all the lipoproteins are really defined by their physicochemical properties, not by their chemical properties, and as we are now beginning to find out, within the LDL pool you have some LDLs that are lighter or heavier, with more or less cholesterol ester, and that may be the determining factor rather than merely the amount of LDL.

Shepherd Perhaps, I could make a statement about the apparent dichotomy between what Professor Kritchevsky says and what I say. There is not actually, as far as I can see, any difference in what we said. When we fed a high cholesterol diet to our subjects, we saw an increment in cholesterol over the course of about a month and then a slow decrement if we continued to feed the cholesterol. So what is happening? The body is adapting to this increased load of cholesterol and if we were to measure cholesterol say 6 months later it is quite likely these individuals would have come back to the basal value. What would be happening under these circumstances is that one is altering the metabolism of the cholesterol itself. So that perhaps the requisite amounts are not being funnelled into these so-called receptor pathways, which are putatively anti-atherogenic, and the other pathways, which are pro-atherogenic.

Cannon, London I would like to challenge the conclusions Professor Kritchevsky and others with similar views draw from the data they cite. We, in Western countries, are not adapted to a high fat/saturated fat diet. Physiologically the human being is adapted to a diet containing maybe between 15% and 30% total fat, little of which is saturated fat. If an entire population – like those of America and Britain – is on an unphysiologically high total fat and saturated fat diet, then those who are born vulnerable to such artificially high levels will eventually die from diseases provoked by too much fat; those who are born invulnerable to the ill effects of fat will not be affected. People can be vulnerable to high fat consumption either for genetic reasons or else for environmental reasons, such as the people of Scotland whose very low consumption of antioxidants in fresh vegetables further increases their risk of heart attacks (unlike the people of France whose high antioxidant intake is protective). So within a population whose average fat intake is way above a physiological amount, you would not expect to obtain a significant correlation between fat intake and cardiovascular incidence. The valid correlations are not within populations but between populations or else within a population, but looking at data over a long period of time.

Kritchevsky The point about people within the same population may highlight the susceptible versus the non-susceptible. As far as what man is adapted to, I am not sure anybody knows. The hunter/gatherers knew how to live but they did not live long enough to find out. The World Health Organization publishes a map of the world every year in which every country is rendered in one of five colours, the colour relating to life expectancy. The countries with the bad diets are the ones that have the high life expectancy. I think that part of what we are seeing is a concommitant of just living longer. There is no question that there are people who are susceptible that have to be sought out. At the moment they are being sought out by their lipid and lipoprotein levels, which only tell part of the story.

Cannon The point applies also to sugar and to fat, as well as to fat and saturated fat. It is a general point.

Kritchevsky It is the old gag – if you stay away from everything you like to eat, women, whisky and late hours you don't live forever. It just seems that way.

Moynahan, London Apropros adaptation to changing circumstances, towards the end of 13th century the climate in Western Europe underwent a remarkable change. That led to famine because of dramatic harvest failures in successive years. It is likely that those who could lay down and store fat, and so on, would have a better chance of surviving to contribute their genes to the following generations. It may well be that we are now paying the price for the survival of our ancestors because now we consume too much food.

Kritchevsky It is also possible that we are paying the price of not overeating but underworking. Because the energy equation is a total balance of not only the amount that is taken in but the amount that is expended. In most developed countries very few people work hard any more. It has been shown in three different surveys in the US that men who work at hard labour are at a much lower risk of colon cancer than are men who have sedentary jobs. So I think we haven't really paid enough attention to that end of the equation.

Shepherd I think to be very cruel about the whole thing you could say that coronary heart disease is of no consequence because it does not limit reproduction, and that may be what happened in the 14th century. People were only living to about the age of 40.

Bender, London When the Finns went on the Italian diet their total energy intake was about 2000 calories less. What would happen to the lipid pattern if you simply reduced the calorie intake without changing the composition of the diet?

Shepherd There have been studies which I think maybe you know more about than I do, but you do alter plasma lipid levels by altering calorie intake.

Kritchevsky Everything in a diet is important, I do not think you can single out any one component. In animal experiments we find that animal protein is more cholesterolaemic than is vegetable protein. Interestingly, if we feed a diet containing animal and vegetable protein in a ratio of 1:1 it has the same effect as feeding the vegetable protein. Everything is important and I think at the present time we just have not reached all the answers. The difficulty comes when people decide yes we now know it all. And we do not.

Wharton, Glasgow I wonder if you could ask the speakers to speculate about whether it is justified to try to lower the plasma cholesterol level in the whole population, bearing in mind the cancer story in the the Paisley study. The message generally used to be yes, but there are now sufficient doubts for this question to be raised.

Kritchevsky I would say that the data suggest, and have suggested for a long time, that at a cholesterol level of over about 240 mg/100 ml or about 6.2 mm/l the danger for CHD begins, but it is still affected by family history and everything else. Dr Michael DeBakey makes some people in the US upset when he says that half of his patients have normal cholesterol. The emerging cancer data are showing that maybe too low is not good either. We just do not know the exact value, although cholesterol is still one of the major risk factors, but not to be seen in isolation. My advice would be above 240 is probably where you have to start looking, but if you have a family history you should be careful all the time.

Wharton But that is in the individual. I was asking about the population policy where you are going to try to alter the plasma distribution of cholesterol in the whole population. At the top end you have problems of cardiovascular death and at the bottom cancer.

Kritchevsky I do not think treating the whole population is a good idea, I think people should be treated individually.

Shepherd I think, at the end of the day, although it is ideal to treat individuals, when it comes to a situation as we have in this country where mortality is amongst the highest in the world and lipid levels are unusually high in comparison to countries where mortality is lower, then we have to adopt not only an individual strategy but we have to have some kind of population strategy as well. Now if you were to take a whole herd of individuals in the UK and lower their cholesterol by 5 mg/dl you would have a major impact on the population but virtually no impact on the individual. I think it is worthwhile making that distinction. I think it is also important to remember that although coronary heart disease does have a serious predation on a population it is changing and therefore there must be something happening that is making it change. It does not change spontaneously for no reason whatsoever. I think cholesterol is one of the factors which does relate to that change. I do not think it is the only factor by any means, but it is worthwhile lowering cholesterol on a national basis.

MacDonald What advice would you give to an individual whose serum cholesterol was very low and he appeared to be fit. Would you tell him to eat saturated fat?

Shepherd I would tell him to go away and enjoy himself.

MacDonald But he might be harbouring a cancer.

Shepherd I remember something else about the cancer story. The Paisley story was one of 15 studies, the others have not shown any relationship between cholesterol and cancer. I think we have got to be careful not to select convenient pieces of data useful for our case.

Kritchevsky A few years ago McMichael in Australia published a review of all the known studies. He felt that after he took out all the caveats, for example people who might have had cancer before entering the trial, there still were concerns. You are right that you pick whatever studies suit your hypothesis. But the problem in this field is that we all have the same data. Everybody just thaws it over the fire of a different prejudice and comes out with a different answer. Somehow we have to get an answer that is general because then all the data may fit it.

Eastwood, Edinburgh Do you think there is anything to be learnt from the different coronary artery disease rates in France and Japan and Scotland and Finland?

Kritchevsky Figures of risk and serum cholesterol show all the countries with their serum cholesterols in a straight line. Yet one point had five countries lined up with the same level of serum cholesterol and vastly different rates of coronary atherosclerosis. Again we get back to the problem that there are still factors that are unknown. Unfortunately there is so much interest in looking at just cholesterol that others are not looked at; partly because of lack of support.

Shepherd Looking at different countries helps you learn how little you know about the situation. If you look at the population of Japan, for example, they have low cholesterol values in relation to us, they live longer than we do but the life expectation differential is actually small. So that we should not be expecting if we lower the population cholesterol level in this country to live another twenty years. We might add if anything another couple of years. What we should be trying to do is prevent people from dying prematurely. That is is the major issue. When it comes to France that is completely enigmatic. Even a Frenchman who is an epidemiologist will tell you that he does not understand the results of their data collection. I think that may be because you are looking at two countries in one, a northern European country and a mediterranean country. I do not know whether there are differentials between the south and the north of France in relation to mortality/ morbidity factors and their diet.

Coubrough, Aberdeen Just a point of information, particularly to the Scots in the audience. The Scottish Home and Health department had a working party looking at prevention of coronary heart disease in Scotland for the last 18 months. Last week it published its report which is available from HM Stationery Office.

Garrow, London We have been presented again with this fascinating contradiction between experimental studies in which changes in dietary fat or P/S ratio or whatever lowered serum cholesterol concentrations, and field studies in which attempts to find a characteristic difference between people of different lipids failed to show this difference. Professor Kritchevsky suggested that this was because there were many genetic and other confounders which obscured the effect in field populations. The other possibility is that the diet surveys were not actually terribly accurate tools. It is very difficult because when you do not find differences in diet between two populations that does not mean the differences do not exist, it just means you have not found them.

Kritchevsky I think you are absolutely right. I think there are a couple of problems. One of them is that when we ask people what they are eating it is usually in terms of fat, carbohydrate, protein and lately fibre. You do not ask about condiments or anything else. For instance, we found that curry powder binds about 20% as much bile acid as does cholestyramine. And curry chicken is a lot better than cholestyramine chicken. There is also this phenomenon of the wish list, which is when you ask somebody what they are eating, they tell what they wish they had eaten rather than what they actually had. I think the only way to really know what people eat is to sit in their lap and measure it.

Shepherd Another thing is that if you continue to measure lipid levels in these individuals who are taken on an experimental basis, locked in a room or whatever, and fed an appropriate diet you will see the cholesterol level following implementation or supplementation of cholesterol in the diet will rise progressively and then will tail off. So you do get adaptation and perhaps the adaptation was complete in those circumstances.

Kritchevsky Or else maybe we just do not follow them long enough.

Passmore, Edinburgh Professor Kritchevsky gave a quotation suggesting that Ancel Keys had lost some of his enthusiasm for cholesterol. Ansell Keys did an enormous amount to generate the whole of the cholesterol and fat story and he remained an enthusiast for a long time. On many occasions I sat on international committees with him, but I have not seen him since he has been in Italy for ten years. I wonder whether he would like to attribute this loss of enthusiasm to either the Italian sun or a good Italian wine, or just perhaps to that growth in wisdom that comes naturally to some of us as we get older.

Kritchevsky Apparently the statement that he made was in response to a question about the cholesterol screening programme. I do not think he has really lost his enthusiasm.

Shepherd I think we should remember what he is saying. He is not saying the differences in plasma cholesterol between individuals do not matter, but that differences in dietary cholesterol intake probably make little change in your blood cholesterol. What is important is differences in dietary saturated fat intake and polyunsaturated fat intake.

Kritchevsky In his early work he stated that it was fat and nothing else and later he decided maybe cholesterol in the diet was also important. When he first started up his whole emphasis was just on fat.

MacDonald Just one or two points that still worry me. I am worried that we are getting into the realms of saying it will be unethical to try and lower serum cholesterol. I think this is a possibility we may have to face. And secondly, I have not really got a consensus view from these distinguished speakers as to whether the cholesterol in the diet is relevant or not.

Kritchevsky I think we do agree that it is not as relevant as the type of fat.

PART TWO

Sugar

5 Sugars in human disease: a review of the evidence

K.W. Heaton

5.1 INTRODUCTION

The correct answer to a question depends on the phrasing of the question. When official bodies have reported on sugars and human disease they have always set up questions in the format of 'Is such and such a disease caused by eating sugar?'. The published evidence is reviewed and found to be insufficient to warrant a firm answer; the conclusion is 'no clear evidence'; and 'no clear evidence' becomes interpreted as plain 'no'. This happens because sugar consumption is already widespread and so it is given the benefit of the doubt. In the language of the American Food and Drugs Administration, it is a GRAS product (generally regarded as safe). In other words it is innocent until proved guilty.

The situation would be quite different if sugars were being introduced into the human diet for the first time. Sugars are essentially food additives, that is, artifical sweeteners, and the rules for introducing a new food additive are very strict. Like a new drug, a new additive is guilty until proved innocent and has to prove its safety in thousands of experiments. Some of these involve feeding it to animals for the whole of their lives in doses several times larger on a body weight basis than human beings are likely to take. If there is any disturbance of metabolism, any shortening of the animal's life and, particularly, if there is any increased incidence of cancer, the additive or drug is refused a licence. It is banned as dangerous. I suggest that, if sugars were subjected to these tests, they would fail all of them. They would also fail on another score, namely, that they increase the likelihood of the main nutritional disease of our civilization, obesity.

Why do sugars have such a privileged position? The vested interests of the powerful food industry and people's reluctance to condemn something they like are only part of the story. There are several other factors: (1) the amount of medical research done on sugars is small compared with that on fats; (2) much of the research is of doubtful relevance, being done on animals or with inappropriate design, e.g. isocaloric exchange of sugars for starches; (3) there are enormous difficulties in proving that a dietary habit has caused a disease which is slowly developing or has a long latent interval of unknown duration. For example, we can only guess when in a person's life a dietary factor would have been operative. We have no way of making an accurate assessment of someone's past diet. If we study a person's diet now and monitor his health for the next 20 years his diet will probably change.

We may never obtain proof of the links between diet and chronic disease. The best we can do is draw up a balance sheet, taking evidence from all relevant sources, and arrive at a probability value.

5.1.1 Identity of the accused – extracellular sugars

A crucial first step is to define a suspect food or eating habit as precisely as possible. By sugars I mean mono- and disaccharides. In nature, sugars are always synthesized within cells and, with the special exceptions of milk and honey, this is where they remain unless man decides to extract them. Milk and honey are unimportant parts of the adult human diet and, for practical purposes, extracellular sugars were not available on a regular basis till the invention of the sugar-cane press and sugar refinery, to be followed by the sugar-beet-extracting plant. Only in the mid-19th century did cane and beet sugars become cheap enough to take their present place as major constituents of the diet. They are now used so extensively in processed foods and drinks that it is easy to think of them as inevitable or normal dietary constituents. They are not. Like all other artifical sweeteners, they are necessary neither for good nutrition nor for good gastronomy.

5.2 PHYSICAL FACTS AND PHYSIOLOGY

The essential facts are that sugars are soluble, they convey calories (energy) and they taste sweet. There are several different sugars in human diets but the major one is sucrose (the disaccharide of glucose and fructose). However, in nutritional terms, the chemical type of sugar is relatively unimportant. What matters is its physical state, that is, whether it is inside or outside cells when it is taken into the mouth. Intracellular sugars are not blamed for any human disease except rare instances of congenital enzyme deficiency and, perhaps, a few cases of food allergy.

Extracellular sugars comprise chiefly cane sugar and beet sugar, but recently glucose syrups like hydrolysed corn starch and now synthetic fructose are being used by some food and drink manufacturers. Extracellular sugars have often been called refined sugars, but this is misleading because what sugar manufacturers call unrefined sugars (dark brown, muscovado, etc.) are also extracellular and, gramme for gramme, it is unlikely that their effects are different from those of white sugar.

Extracellular sugars have also been called fibre-depleted sugars. It is quite a good term but only hints at the essential point, namely, that the sugars are unpackaged and exposed, ready for immediate action. This action may be to feed bacteria in dental plaque or to diffuse to the intestinal mucosa, where the sugars stimulate the release of hormones and are absorbed into the bloodstream.

5.2.1 Packaging and some effects of losing it

When we swallow **intra**cellular sugars they are not contained within single, isolated cells but in large clumps of cells, that is, in partly chewed lumps of fruit or root. This is important because the rate at which food leaves the stomach depends on the particle size of the food (large particles being held back until they have been reduced to 1–2 mm by the churning action of the gastric antrum) and the rate at which food leaves the stomach determines how quickly it is digested and absorbed and, therefore, how much insulin is secreted in response to the meal (Jonderko, 1989). In other words, physical intactness or histological integrity of a food slows down its digestion and moderates the insulin response to it. As an illustration, the peak insulin level after a meal of apples was 23.9 mU/1, but after an equivalent amount of apple puree it was 32.2 and after apple juice it was 44.7 (Haber *et al.*, 1977).

The physical form of a food also determines how easily it is ingested. Intact food, like a raw apple, is solid and has to be chewed. Chewing slows down the act of eating and makes it more laborious. When food is harder to eat, less is eaten (Heaton, 1980). It follows that when food is made easier to eat, other things being equal, more will be eaten.

Production of extracellular sugars is the prime example of food being made easier to eat. As Cleave (1974) pointed out, who would eat $1\frac{1}{2}$ lb (700 g) of sugar beet every day? Yet that is equivalent to consuming 100 g sugar, which is roughly the average daily consumption of men in this country. Of course, nobody eats sugar beet, but a realistic comparison is with $1\frac{3}{4}$ lb (800 g) of apples, equivalent to seven medium-sized apples. In a direct comparison of apples with apple juice (which can be considered a proxy for extracellular sugars in general), the time needed to consume the juice was reduced from 16 to 2 minutes and the amount of satiety (feeling of fullness) generated was much less, so that the subject felt ready to eat again within an hour instead of two hours (Haber *et al.*, 1977). This and other studies comparing fruit and fruit juice (Bolton *et al.*, 1981) suggest that taking sugars in extracellular form is likely to encourage increased calorie intake over that derived from intact whole foods. But do sugars actually have this effect?

5.3 SUGARS AND SPONTANEOUS CALORIE INTAKE

5.3.1 Effects of decreasing sugar consumption

Two types of study have been reported.

(a) Elimination of extracellular sugars from the diet

In this type of study, people who normally take sugars regularly are asked to leave them out of their diet as far as they can and instead to eat any sugar-free foods they fancy. The effect on calorie intake is judged by the effect on body weight and, if the subjects can be persuaded to record their food and drink intake, by dietary records. Five studies have recorded weight change under these circumstances and, in all five, body weight fell (Table 5.1).

This is impressive because in two of the studies the subjects were told not to lose weight and in two others they were told not to go hungry. In two studies calorie intake was assessed and this fell on sugar exclusion compared with a diet in which sugars were encouraged.

Sceptics have disparaged these findings by suggesting that any drastic restriction of food choice, indeed any major involuntary change in diet, will make people eat less and reduce their calories. This claim is not borne out by the evidence. In one of the above studies a group of people were asked to cut down on starch; they did not lose weight, in fact they gained a little despite stating they found it difficult to cut down starch (Mann *et al.*, 1970). In two other studies the subjects were asked, in another experimental period, to increase their intake of sugars and sugar-containing foods and to limit their intake of fruit and vegetables. With this unwanted change in diet they *gained* weight (Thornton *et al.*, 1983; Werner *et al.*, 1984). Holidays involve a major change in diet, but few people lose weight on their holidays!

Table 5.1 Unintentional weight loss in five studies involving reduction of extracellular sucrose intake

Reduction in sucrose intake (g/day)	*Article of diet replacing sucrose*	*Duration of low sugar intake (days)*	*Mean weight loss (lb)*	*Reference*
73	Starchy foods	140	5*	Mann *et al.* (1970)
Not stated	Any other food	70	2.3	Rifkind *et al.* (1966)
220	Any other food	12	2	Porikos *et al.* (1982)
100	Mainly fruit and vegetables	42	3.5	Thornton *et al.* (1983)
96	Fruit and vegetables	42	3	Werner *et al.* (1984)

* Actually 2.8 lb, but two control groups each gained 2.0 lb

A more valid criticism of the above studies is that, because many people believe in the sugar–overweight connection, some of the volunteers may have expected to lose weight and have reduced their food intake to make sure they did. They may even have used the experiment as a chance to slim. Unlikely though this is, it cannot be refuted in studies of free-living people buying or choosing their own foods. This criticism is, however, refuted by the second type of study.

(b) Covert substitution of aspartame for sugars

In this type of study food intake is measured precisely over a period of days by keeping volunteers in a closed environment and providing them with generous helpings of the foods and drinks they like, including sugary ones, measuring how much they leave uneaten. After a few days the sweet foods and drinks are changed without the volunteers' knowledge, so that they are sweetened with aspartame instead of sugar, or vice versa. This type of research is feasible only for short periods in small numbers of people. Nevertheless, the results have been clear-cut. When eight obese people who normally took a quarter of their calories as sucrose were studied in this way, their energy intake fell by 23% in the first 3 days of covert aspartame substitution and stayed reduced by 14% in the next 3 days (Porikos *et al.*, 1977). In a second study, six normal-weight men unconsciously reduced their calories by 24% in the first 3 days of aspartame substitution and by 15% in the next 9 days (Porikos *et al.*, 1982).

The message of these elegant experiments is that human beings are bad at detecting the fact that sugar conveys calories as well as a sweet taste. Alternatively, we cannot or will not reduce our intake of other foods to allow for it. In a sense, this is no surprise. Sucrose has long been stigmatized as empty calories and these experiments show it is no empty gibe. But it is an important concept that sugars are essentially calorie-rich sweeteners. This concept puts sugar back into its ancient historical context of an expensive condiment which was used sparingly, like salt.

The logical conclusion from all the studies I have reviewed, including those comparing fruits and fruit juices, is that consumption of extracellular sugars unwittingly and inevitably inflates energy intake above what it would be if such sugars were not available. In brief, they cause overnutrition.

Of course, sugar consumption is not the only cause of overnutrition but it is a ubiquitous one. If sugars supply 16% of the calories available for human consumption in the UK, as is officially stated (Central Statistical Office, 1989), one can infer that the average Briton may be obtaining up to 16% more calories than he needs.

Since most people in Britain eat extracellular sugars, most of us must be overnourished. This being so, most of us must be overweight or gaining weight, or we must have taken steps to control our weight. Is this so?

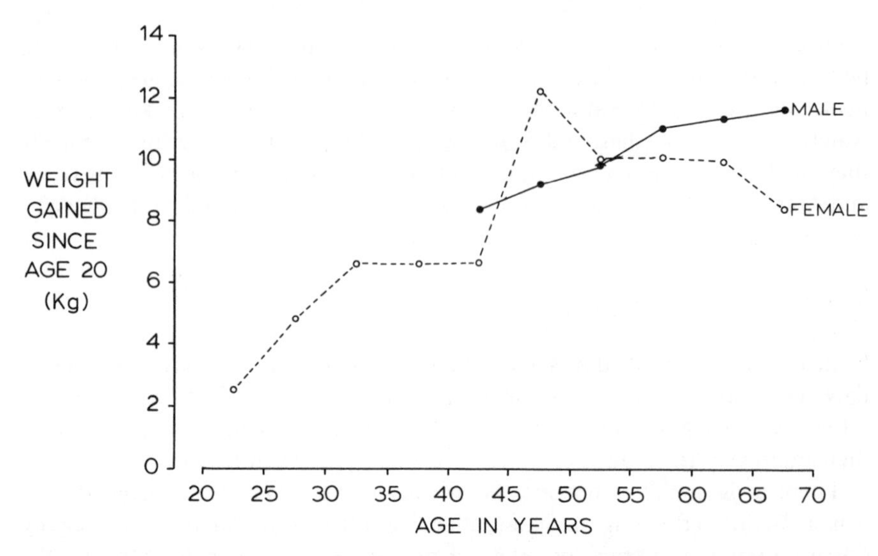

Figure 5.1 Difference between current weight and weight at age 20 in healthy men and women of different ages (unpublished data from the East Bristol Gallstone Survey, Heaton *et al.*, 1990)

5.3.2 Weight gain in British adults

Several studies have examined the change in weight of the population since early adulthood and all have found that it goes up. Durnin (1968) followed 611 male students for 15 years and found they had gained, on average, 8.5 kg. Only 15% stayed within 2.3 kg of their initial weight. As part of an epidemiological survey in Bristol we have weighed people and also asked them what they weighed at age 20. Most people remember very well. Eighty-five per cent of people had gained weight. Weight gain rose with age in men and peaked at 50 years in women (Fig. 5.1). This is exactly what one would expect to find if most people in the population are regularly consuming more calories than they need.

The ubiquity of overnutrition matters because some of the commonest and most important diseases of our society are linked with overnutrition. There are social implications too. The faster a girl gains weight, the earlier she has her menarche (Garn *et al.*, 1986).

5.4 HEALTH CONSEQUENCES OF OVERNUTRITION

5.4.1 Longevity

For obvious reasons it is impossible to obtain unequivocal information on the relationship between dietary habits and longevity in man. However, it has been known for many years that restraining laboratory rats from eating their pelleted, artifical diet *ad libitum* extends their life span (McCay *et al.*, 1939). In man, life expectancy varies with body fatness. In 1959, 755 502 healthy Americans were weighed and measured and then their state of health and, if they died, their cause of death was determined every year for 13 years (Lew and Garfinkel, 1979). Compared with people of average weight, those who were overweight were more likely to die for any reason, but particularly from coronary heart disease, cancer, diabetes and stroke. In men, heart disease was lowest in the underweight. A report from the Metropolitan Life Insurance Company of New York showed that, when fat people were provoked by a loaded premium to shed enough weight to qualify for a standard premium, they restored their life expectancy to normal or near normal (Dublin, 1953). Of course, they may have made other life-style changes which contributed to their improved life expectancy.

There is debate as to whether thinness is an advantage. Some data suggest it is a disadvantage, but this may reflect the effects of smoking, pre-existing disease and other adverse factors associated with low weight.

5.4.2 Cancer

Several cancers are commoner in obese people (Lew and Garfinkel, 1979). They are cancer of the colon, breast, uterus (endometrium), ovary, prostate and gallbladder. Between them they account for a large proportion of cancer deaths in this country. It has long been known among workers with experimental cancer in animals that to obtain the maximum yield of cancers the diet must be semisynthetic (which means it contains sugars) and the animals must be given the diet *ad libitum* (Tannenbaum, 1942; Tannenbaum and Silverstone 1957). Restricting calories by 20–30% dramatically reduces the susceptibility of animals to many cancers. It is often said to be impracticable for ordinary people to cut down their food intake by 20–30%. In fact, calorie intake was curbed by just this amount when people cut down their consumption of sugars from 110 to 10 g daily but otherwise ate and drank freely (Thornton *et al.*, 1983; Werner *et al.*, 1984).

Very few people have looked for an association between cancer and sugar consumption in man. For reasons stated earlier, the odds are against finding any such association even if it exists. Nevertheless, Bristol *et al.* (1985) reported that patients with colorectal cancer had been in the habit of eating more sugar (and fat) than healthy controls. The biggest difference between the two groups was in sugar-fat combinations (cakes, biscuits, puddings, chocolate, etc.). Of course it cannot be proved that this was the diet the patients were eating at the time they developed cancer. It may be that a hypercaloric diet encourages the growth of cancers rather than their initiation.

Table 5.2 Disorders associated with obesity

Hypertension*	Osteoarthritis
Hyperlipidaemias*	Gout
Diabetes*	Renal stones
Gallstones*	Varicose veins
Gastro-oesophageal reflux	Intertrigo
Hernias	Accidents

* Especially associated with truncal/central obesity

In addition to the Bristol study there have been studies in America (Jain *et al.*, 1980; Lyon *et al.*, 1987) and Australia (Potter and McMichael, 1986) which have shown patients with colorectal cancer to be high calorie consumers but, for some reason, the French behave differently (Macquart-Moulin *et al.*, 1983; Berta *et al.*, 1985).

5.4.3 Diseases associated with obesity

Many diseases are generally believed to be commoner in obese people (Table 5.2). To the extent that obesity is caused by consuming sugars, these diseases may, in part be attributable to sugar consumption. Such an idea would be more plausible if there was overlap between the metabolic disturbances associated with obesity and the metabolic effects of consuming sugars.

Metabolic disturbances associated with obesity

Obesity is a dynamic, metabolically-active condition. It is associated with undesirable changes in plasma levels of triglycerides, HDL cholesterol and insulin. These are undesirable because they are believed to be risk factors (or, at least, risk indicators) for several common diseases – coronary heart disease, hypertension, type 2 diabetes, gallstones, gout and renal stones (Table 5.3). There is much overlap between the three metabolic changes; people who have one tend to have the others too and there is similar overlap between the diseases associated with these metabolic changes.

These metabolic disturbances sometimes exist in non-obese people, when their cause is uncertain (Ruderman *et al.*, 1981). The point to note is that all of them can be caused by a high intake of sugars.

5.5 METABOLIC EFFECTS OF DIETARY SUGARS

Most studies of the metabolic effects of sugars have consisted of experiments in which volunteers had a high sugar intake in one period and a low intake in

Table 5.3 Diseases believed to be associated with fasting hypertriglyceridaemia, low plasma HDL cholesterol and high plasma insulin

Hypertriglyceridaemia	*Low HDL cholesterol*	*Hyperinsulinaemia*
Coronary heart disease	Coronary heart disease	Coronary heart disease
Diabetes	Gallstones	Hypertension
Gallstones		Diabetes
Gout		Gallstones
Renal stones		Renal stones

the other, the total calorie intake being kept constant by raising the intake of starch. Such a study design has the appearance of good science, since it controls for a non-sugar variable, namely calories. However, calories are the very essence of sugars and to control for them is artificial. One is left simply comparing two chemical forms of carbohydrate. Despite this handicap, the extensive studies of Reiser and his colleagues at Baltimore have shown that a high intake of sucrose, that is 18–33% of calories isocalorically exchanged for starch, has the following adverse effects (Reiser *et al.*, 1978, 1981a,b, 1986; Israel *et al.*, 1983):

- increased fasting plasma triglycerides in men and in post-menopausal women;
- decreased HDL cholesterol (but increased VLDL and LDL cholesterol);
- increased fasting insulin, greater in men than women;
- increased fasting glucose;
- increased insulin response to a sucrose load;
- increased uric acid.

Some of these changes occurred only in the 15% of the population who are most susceptible to the effects of sugars and critics have belittled Reiser's work on these grounds. However, 15% is a substantial minority and who is to know if he is amongst them? Another criticism of Reiser's work is that his subjects had to consume 75% of their daily calories in their evening meal.

Against these limitations must be set the fact that Reiser controlled for the calorie-inflating effect of sucrose which may have reduced its biochemical effects. In line with this, Werner *et al.* (1984) found the same effects of sucrose as Reiser in respect of triglycerides and HDL cholesterol, even though they did not select their subjects for carbohydrate sensitivity nor enforce any particular eating pattern. All the same, Werner's volunteers were selected in that they were people with cholesterol-rich gallstones and studies are needed on a large, representative sample of the population.

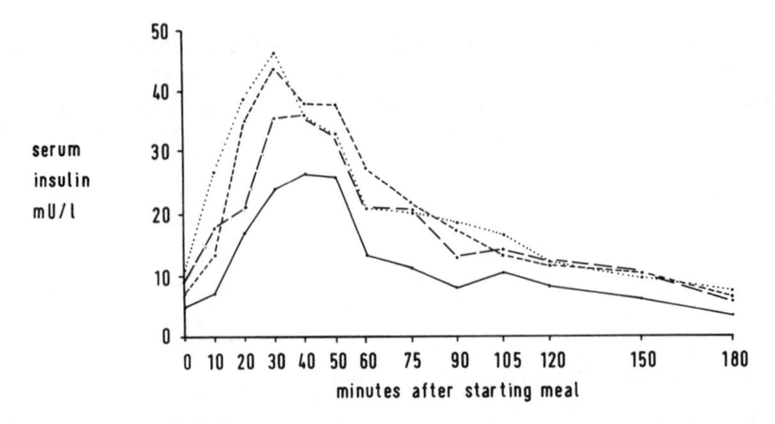

Figure 5.2 Plasma insulin response to four test meals with similar nutrient composition. Mean data from nine healthy volunteers (Oettlé *et al.*, 1987). Key:---- mars, tea; ...... coca cola, crisps; —— raisins, peanuts, tea; —·— bananas, peanuts, tea.

5.5.1 Hyperinsulinaemia and the role of sugars

There is mounting evidence that hyperinsulinaemia is a risk factor for several important diseases, including coronary heart disease (Lichtenstein *et al.*, 1988). The link with heart disease may be indirect, because insulin tends to vary inversely with HDL cholesterol levels (Jarrett, 1988), or insulin may have a direct effect by stimulating cholesterol synthesis (Stout, 1987) and by raising the blood pressure (Reaven and Hoffman, 1987). If it has a direct effect, then any dietary measure which tends to raise plasma insulin must be suspected of contributing to heart disease. In addition, there is suggestive evidence that high insulin levels increase the risk of gallstones (Scragg *et al.*, 1984) and kidney stones (Rao *et al.*, 1982).

Sugars can increase plasma insulin levels in several ways: (1) by amplifying the insulin response to a meal or drink; (2) by inducing obesity, which makes the body tissues resistant to the action of insulin and is compensated by the secretion of more insulin; (3) by inducing insulin resistance directly, as found by Reiser *et al.* (1981a, 1986).

The tendency of extracellular sugars to induce excessive insulin responses to meals was shown by comparing apple juice and orange juice with their respective whole fruits (Haber *et al.*, 1977; Bolton *et al.*, 1981). It has also been shown by studies of realistic snacks containing extracellular sugars, namely a Mars bar and a cup of tea and a can of Coca-Cola with a bag of crisps (Oettlé *et al.*, 1987). Normal volunteers reacted to these snacks with a significantly greater output of insulin than they had after a wholefood snack containing intracellular sugars, namely raisins and peanuts (Fig. 5.2). Adding a can of Coca-Cola to a meal has the effect of amplifying the insulin response to that meal (Mazzaferri *et al.*, 1984). It should, however, be noted that very ripe bananas and grapes are not too different in their effects from extracellular sugars (Bolton *et al.*, 1981; Oettlé *et al.*, 1987).

5.6 SUGARS AND CROHN'S DISEASE

Crohn's disease is a rare but sometimes distressing disease of the intestines which become chronically inflamed and narrowed. It often requires surgical treatment but there is no cure. It seems to be a disease of 20th century Western civilization, especially in northern Europe and North America. Its cause is unknown. However, there is one clue. When groups of patients have been questioned about their diet they have invariably been found to consume excessive amounts of sucrose compared with healthy controls. This has been found in no less than 17 studies and must hold a clue to the cause of the disease (Heaton, 1988). Of the various possible links perhaps the most promising is the effect of sugars on the permeability of the intestinal wall. When taken in concentrated form, sugars make the intestine leaky and this might allow damaging substances to be absorbed into the intestinal wall, setting up an inflammatory defence reaction.

5.7 SUMMARY AND CONCLUSIONS

Extracellular sugars, by their very nature, are ingested and absorbed abnormally easily. This has been shown experimentally to inflate calorie intakes and insulin levels in the blood. In the long term this is bound to cause weight gain and probably induces metabolic disturbances (hypertriglyceridaemia, hyperinsulinaemia, low plasma HDL cholesterol). All this favours the development of coronary heart disease, diabetes, gallstones and several common cancers. It will never be possible to prove that sugars cause any of these diseases, nor is it likely that sugars are the only cause of any of them. However, there are no nutritional or medical benefits from sucrose, or extracellular sugars in general. Consumption of sugars in their natural intracellular state, in fruit and vegetables, allows sugars into the body in amounts and at rates to which the body is adapted and ensures that the body receives minerals and vitamins as well as calories.

REFERENCES

Berta, J.-L., Coste, T., Rautureau, J., *et al.* (1985) Alimentation et cancers recto-coliques. Résultats d'une étude 'cas-témoin'. *Gastroenterol. Clin. Biol.*, **9**, 348–53.

Bolton, R.P., Heaton, K.W. and Burroughs L.F. (1981) The role of dietary fiber in satiety, glucose, and insulin: studies with fruit and fruit juice. *Am. J. Clin. Nutr.*, **34**, 211–17.

Bristol, J.B., Emmett, P. M., Heaton, K.W. and Williamson, R.C.N. (1985) Sugar, fat, and the risk of colorectal cancer. *Br. Med. J.*, **291**, 1467–70.

Central Statistical Office (1989) *Annual Abstracts of Statistics*, HM Stationery Office, London.

Cleave, T.L. (1974) *The Saccharine Disease*, John Wright, Bristol.

Dublin, L.I. (1953) Relation of obesity to longevity. *New Engl. J. Med.*, **248**, 971–4.

Durnin, J.V.G.A. (1968) The change in body weight of a young adult human population after an interval of 15 years. *J. Physiol.*, **198**, 22P.

Garn, S.M., LaVelle, M., Rosenberg, K.R. and Hawthorne, V.M. (1986) Maturational timing as a factor in female fatness and obesity. *Am. J. Clin. Nutr.*, **43**, 879–83.

Haber, G.B., Heaton, K.W., Murphy, D. and Burroughs, L. (1977) Depletion and disruption of dietary fibre. Effects on satiety, plasma-glucose, and serum-insulin. *Lancet*, **ii**, 679–82.

Heaton, K.W. (1980) Food intake regulation and fiber, in *Medical Aspects of Dietary Fiber* (eds G.A. Spiller, and R.M. Kay), Plenum, New York, pp. 223–38.

Heaton, K.W. (1988) Dietary sugar and Crohn's disease. *Can. J. Gastroenterol.*, **2**, 41–4.

Israel, K.D., Michaelis, O.E., Reiser, S. and Keeney, M. (1983) Serum uric acid, inorganic phosphorus, and glutamic-oxalacetic transaminase and blood pressure in carbohydrate-sensitive adults consuming three different levels of sucrose. *Ann. Nutr. Med.*, **27**, 425–35.

Jain, M., Cook, G.M., Davis, F.G., *et al.* (1980) A case-control study of diet and colo-rectal cancer. *Int. J. Cancer*, **26**, 757–68.

Jarrett, R.J. (1988) Is insulin atherogenic? *Diabetologia*, **31**, 71–5.

Jonderko, K. (1989) Effect of calcitonin on gastric emptying in patients with an active duodenal ulcer. *Gut*, **30**, 430–5.

Lew, E.A. and Garfinkel, L. (1979) Variations in mortality by weight among 750 000 men and women. *J. Chron. Dis.*, **32**, 563–76.

Lichtenstein, M.J., Yarnell, J.W.G., Elwood, P.C., *et al.* (1988) Sex hormones, insulin, lipids, and prevalent ischemic heart disease. *Am. J. Epidemiol.*, **126**, 647–57.

Lyon, J.L., Mahoney, A.W., West, D.W., *et al.* (1987) Energy intake: its relationship to colon-cancer risk. *J. Natl Cancer Inst.*, **78**, 853–61.

McCay, C.M., Maynard, L.A., Sperling, G. and Barnes, L.L. (1939) Retarded growth, life span, ultimate body size and age changes in the albino rat after feeding diets restricted in calories. *J. Nutr.*, **18**, 1–13.

Macquart-Moulin, G., Durbec, J-P., Cornée, J., *et al.* (1983) Alimentation et cancer recto-colique. *Gastroenterol. Clin. Biol.*, **7**, 277–86.

Mann, J.I., Truswell, A.S., Hendricks, D.A. and Manning, E. (1970) Effects on serum-lipids in normal men of reducing dietary sucrose or starch for five months. *Lancet*, **i**, 870–2.

Mazzaferri, E.L., Starich, G.H. and St Jeor, S.T. (1984) Augmented gastric inhibitory polypeptide and insulin responses to a meal after an increase in carbohydrate (sucrose) intake. *J. Clin. Endocr. Metab.*, **58**, 640–5.

Oettlé, G.J., Emmett, P.M. and Heaton, K.W. (1987) Glucose and insulin responses to manufactured and whole-food snacks. *Am. J. Clin. Nutr.*, **45**, 86–91.

Porikos, K.P., Booth, G. and van Itallie, T.B. (1977) Effect of covert nutritive dilution on the spontaneous food intake of obese individuals: a pilot study. *Am. J. Clin. Nutr.*, **30**, 1638–44.

Porikos, K.P., Hesser, M.F. and van Itallie, T.B. (1982) Caloric regulation in normal-weight men maintained on a palatable diet of conventional foods. *Physiol. Behav.*, **29**, 293–300.

Potter, J.D. and McMichael, A.J. (1986) Diet and cancer of the colon and rectum: a case-control study. *J. Natl Cancer Inst.*, **76**, 557–69.

Rao, P.N., Gordon, C., Davies, D. and Blacklock, N.J. (1982) Are stone formers maladapted to refined carbohydrate? *Br. J. Urol.*, **54**, 575–7.

Reaven, G.M. and Hoffman, B.B. (1987) A role for insulin in the aetiology and course of hypertension. *Lancet*, **ii**, 435–7.

Reiser, S., Bohn, E., Hallfrisch, J., *et al.* (1981a) Serum insulin and glucose in hyper-insulinemic subjects fed three different levels of sucrose. *Am. J. Clin. Nutr.*, **34**, 2348–58.

Reiser, S., Bickard, M.C., Hallfrisch, J., *et al.* (1981b) Blood lipids and their distribution in lipoproteins in hyperinsulinemic subjects fed three different levels of sucrose. *J. Nutr.*, **111**, 1045–57.

Reiser, S., Hallfrisch, J., Fields, M., *et al.* (1986) Effects of sugar on indices of glucose tolerance in humans. *Am. J. Clin. Nutr.*, **43**, 151–9.

Reiser, S., Hallfrisch, J., Michaelis, O.E., *et al.* (1978) Isocaloric exchange of dietary starch and sucrose in humans. I Effects on levels of fasting lipids. *Am. J. Clin. Nutr.*, **32**, 1659–69.

Rifkind, B.M., Lawson, D.H. and Gale, M. (1966) Effect of short term sucrose restriction on serum lipid levels. *Lancet*, **ii**, 1379–81.

Ruderman, N.B., Schneider, S.H. and Berchtold, P. (1981) The 'metabolically-obese', normal weight individual. *Am. J. Clin. Nutr.*, **34**, 1617–21.

Scragg, R.K.R., Calvert, G.D. and Oliver, J.R. (1984) Plasma lipids and insulin in gall stone disease: a case-control study. *Br. Med. J.*, **289**, 521–5.

Stout, R.W. (1987) Insulin and atheroma. *Lancet*, **i**, 1077–9.

Tannenbaum, A. (1942) The genesis and growth of tumours. II Effects of calorie restriction per se. *Cancer Res.*, **2**, 460–7.

Tannenbaum, A. and Silverstone, H. (1957) Nutrition and the genesis of tumours, in *Cancer*, Vol. 1. (ed. R.W. Raven), Butterworth, London, pp. 306–34.

Thornton, J.R., Emmett, P.M. and Heaton, K.W. (1983) Diet and gall stones: effects of refined and unrefined carbohydrate diets on bile cholesterol saturation and bile acid metabolism. *Gut*, **24**, 2–6.

Werner, D., Emmett, P.M. and Heaton, K.W. (1984) The effects of dietary sucrose on factors influencing cholesterol gallstone formation. *Gut*, **25**, 269–74.

6 Sugars not for burning

A.A. Jackson

6.1 INTRODUCTION

In the government report on dietary sugars (COMA, 1989), one point of interest made abundantly clear, both from the comments invited from the public at large during the preparation of the report, and the general response since the publication of the report, is the extent to which different individuals are able to interpret the same set of data in completely different ways. This highlights the point made in the report that there are too few good studies to enable reasonable conclusions to be drawn on a number of important points. It is not my intention to discuss the contents of the report, which has been out for a sufficient period of time for all who are interested to have studied it in detail at leisure. Rather, I would prefer to discuss a number of issues of a more general nature, which deal with the way in which we conduct our approach to nutritional enquiry in general, but using sugars as a specific example. Undoubtedly one of the most important factors which is highlighted by the report is that no individual or group can afford to be complacent about the state of our knowledge of the interaction between dietary sugars and either health or disease. From a nutritional point of view our knowledge and understanding is woefully inadequate.

I would like to spend a little bit of time in considering the nature of these inadequacies, because I believe that some of the speculations into which Dr Heaton has been tempted can only be permissible in a situation where we have an inadequate basis of understanding of the requirements for sugars and the extent to which those requirements may or may not be exceeded. In all fairness a debate can only exist in a climate where options exist for differences of interpretation, and hence of opinion. I would not like anyone to misconstrue my comments as representing fundamental disagreement with all the points made by Dr Heaton, nor as an endorsement of what has generally been presented as 'industry's view', for which I hold no brief.

6.2 PRINCIPLES OF NUTRITION

Many of the points which I wish to make are in fact very simple. Some may find them boringly simple and those people are entitled to go to sleep and I shall not be offended. However, I hope that for one or two of you, although the points may be simple, they will be apposite because they will assist in focusing your minds on some of the important concepts of nutrition, which are often overlooked or presented in an unsatisfactory way. It is my belief that in nutrition we do a great disservice to ourselves and to our discipline because we are tempted into reducing all considerations to superficialities. We have a tendency either to overlook or to ignore the fundamentals and do not even take enough care to ensure that we are clear or agreed about these fundamentals. Clearly, in terms of policy there is the need to draw broad principles, but this should not lead us into a state of overconfident belief that these broad principles necessarily stand for all time as representing the true nature of the world. My perspective in this debate is that of a nutritionist, not the perspective of a dietitian. It is not my intention to consider foods at all, if anything, I shall rather focus upon nutrients. Earlier, the discussion developed by Professor James explored the consideration that if we are to become involved in making policy decisions then there is the need for a suitable basis of information which is both reliable and reasonably comprehensive. Any approach which is substantially dependent upon ignorance or guesswork cannot be considered to be acceptable or satisfactory. I would like to recognize this imperative, and to apply it to the basis of information from which we have worked in order to gain an understanding about the nutritional requirement for dietary sugars.

6.2.1 Energy balance

We all view the world from different perspectives and my personal point of departure is exemplified by the concepts of balance as illustrated in Fig. 6.1. For any healthy adult, body weight and body composition remain essentially constant over long periods of time, periods of years in many instances. This must mean that individuals remain in balance for energy and all nutrients. Thus, the daily intake of energy must equate to the daily expenditure of energy. An average adult male with a moderately sedentary lifestyle, will ingest in the order of 10 MJ per day, 2500 kcal per day. In the UK about 45% of the energy will derive from carbohydrate with 40% coming from fat and about 15% coming from proteins (Gregory *et al.*, 1990). Surprisingly, there is only a relatively narrow range of variability in these proportions amongst the population, regardless of the types of foods eaten. The same individual will inevitably expend 10 MJ per day, 2500 kcal per day, to remain in balance. This

DAILY INTAKE	10 MJ 2500 kcal
Carbohydrate	45%
Fat	40%
Protein	15%
DAILY EXPENDITURE	10 MJ 2500 kcal
Carbohydrate	45%
Fat	40%
Protein	15%

Figure 6.1 A normal adult man leading a sedentary lifestyle will ingest about 10 MJ each day and expend an equivalent amount of energy. Both the intake and the expenditure will comprise similar proportions of carbohydrate, fat and protein.

also means that the 10 MJ of energy will inevitably derive from the oxidation of substrate represented as 45% from carbohydrate, 40% from fat and 15% from protein. On a daily basis equivalent amounts of carbohydrates, fats and proteins are oxidized in the body as are taken in the diet. This has to be true, otherwise body weight and body composition could not be maintained with constancy.

Most textbooks of biochemistry, dietetics or nutrition imply that carbohydrates and fats are preferred sources of energy, with protein being 'protected' or 'specially utilized' for tissue synthesis and repair. The implication is given that under normal circumstances proteins are not oxidized as a fuel by the body. Obviously, on a daily basis we oxidize an equivalent amount of protein to that taken in the diet, although the complexities of protein turnover clearly indicate that it is not necessarily the same amino acids as were ingested which are oxidized. Therefore, the suggestion that dietary carbohydrate protects or saves protein has to be seen as being of importance or relevance in only a very limited context related to growth or the repletion of lost tissues. Our understanding of nutritional balance for carbohydrates in general has to be placed against this background.

6.2.2 What does essential mean?

One of the dangers associated with the mistaken appreciation of the relationship between dietary energy and protein which has crept into the literature is that protein has been elevated to an unusual position of importance in nutritional terms over fats and carbohydrates. The prevailing wisdom is presented in a way which suggests that fats and carbohydrates are simply sources of energy, whereas proteins are for body building – implying a difference of objective from the point of view of the body in the way it utilizes the different macronutrients. This is patently not correct. There is an essential requirement by the body for carbohydrate and for fat as well as for protein.

However, the term essential can be used in a confusing way and does present slight difficulties. I had not appreciated just how difficult it could be until I started to give a lecture series to a group of mixed nutrition and biochemistry students in their third undergraduate year. The topic of the lecture was proteins and amino acids and required an appreciation of the concept of essential and non-essential amino acids. As it was a third-year course I assumed that all the students had a prior working knowledge of the concept. One of the more vociferous individuals in the class stopped me in mid-flow to say that they were unable to follow what I was saying through lack of appreciation of the concept of essential amino acids. I was relatively unsympathetic to their difficulties. They responded to my inconsiderate approach by leaving the course, never to return. Slightly crestfallen, I thought I ought to discuss the situation with the Professor of Biochemistry. I told him that I was not able to understand what was wrong with the biochemistry students, as they seemed to be unable to differentiate between essential and non-essential amino acids. His response: 'Well, what is the difference?'. So far as he was concerned, being a microbial biochemist, the concept of an essential amino acid did not exist. There was no need for such an idea because all the microbes with which he dealt were able to make a full array of amino acids as and when they were required. What, therefore do we mean when we are talking about essentiality?

The concept which has grown up is that an essential nutrient is one which has to be provided preformed in the diet because it cannot be made by the body. There is the need to differentiate those nutrients which have to be provided preformed as substrate to the cells of the body for their normal function. These compounds may be made in other parts of the body from either dietary or endogenous precursors. In this sense, an alternative way of looking at the problem is to consider that an essential nutrient is one which is absolutely required by the cell for its normal function. Depending upon whether the problem is considered at the level of the cell or at the level of the whole body, the perception and interpretation might change.

Nobody would deny that most, if not all cells, have an absolute requirement for glucose for a range of synthetic reactions, if not as a source of energy. If this applies at the level of the cell how can it be extrapolated to the level of the whole body and interpreted in a way which takes cognizance of function? Is it reasonable, within the context of whole body metabolism, to perceive of carbohydrates as being essential nutrients? Do carbohydrates have to be provided preformed in the diet? When one tries to put the concept of essentiality into context it is not possible to reduce it to a single statement.

6.2.3 Conflict between synthesis and oxidation: balance and essentiality

When one tries to put the two concepts together, the idea of balance and the idea of nutritional essentiality, one ends up with a conundrum. We have an

Carbohydrate: Essential
Fat: Essential
Protein: Essential

CONFLICTING DEMANDS

Carbohydrate
Fat
Protein

STRUCTURE/SYNTHETIC———OXIDIZE/ENERGY

Figure 6.2 Each of the three macronutrients, carbohydrate, fat and protein, play an essential part in synthetic metabolic reactions. Therefore, there is a conflicting demand upon the dietary intake for synthetic reactions on the one hand and oxidative reactions as a source of energy on the other.

essential requirement for carbohydrate, fats and proteins and yet on a daily basis we oxidize the equivalent of our intake of carbohydrate, fats and protein. The appearence is of conflicting demands upon our macronutrient intake. On the one hand they are required for structural purposes as substrate for synthetic reactions. On the other, they are oxidized as fuels to provide the energy to drive the synthetic reactions of the body (Fig. 6.2). The objective of scientific enquiry is to understand the nature of this dynamic equilibrium and to understand the factors which determine whether an imbalance falls out in one direction or the other at any one point in time, recognizing that over extended periods of time, balance must be maintained between both options. In order to make sense of the relationship an appreciation is required of the role played by turnover within metabolism. Unless the problem is posited within this framework of consideration, it is not possible to provide answers which represent a rational and coherent response to questions about the need for dietary sugars. Clearly the nature of the answer obtained is in large part determined by the nature of the question which is asked. It is my perspective that so far as dietary sugars are concerned we have, to a certain extent, been asking the wrong questions and therefore should not be surprised when we obtain answers which are not altogether satisfactory.

6.3 REQUIREMENTS FOR DIETARY SUGARS

I would like to spend a little time asking those questions, within a perspective of reasonable nutritional considerations, using the standard approach of nutritional enquiry. Nutrition asks:

- What is the minimum amount of substance that is needed in the diet to maintain adequate health and well-being?

- What is the maximum amount of substance in the diet that the body can tolerate without adverse effects?
- What amount of a substance in the diet allows for optimal function?
- Do these amounts change from one circumstance to another, with age or sex, in health and disease?

Invariably, when one attempts to define the requirements, one is faced with a value judgement and the need to answer the question 'Requirements for what metabolic state?'. Depending upon the state the requirement might change. It is, however, possible to start from first principles and ask whether there is an absolute requirement for glucose or sugars in the diet. Is there any evidence either for or against the proposition? The literature does not provide many answers to the question, and there are very few studies in which humans have been fed for extended periods of time on diets which are free of sugars or other carbohydrates.

6.3.1 Minimal requirement for dietary sugars

One approach to the problem is that adopted by MacDonald (1987), who asked whether it was possible to define a metabolic requirement for carbohydrates in general, or sugars specifically. He chose as his outcome the need to inhibit ketosis, that is to enable the oxidation of fatty acids. Wisely cautious, he did not impute any pathology to the ketotic state, but recognized that ketosis might be associated with mild disturbances of cellular function with manifestations such as errors of judgement. He estimates that the minimum requirement to limit ketosis in an adult man would be of the order of 180 g carbohydrate per day. In other words, the body has to produce from endogenous and exogenous sources 180 g glucose. This should allow fat oxidation to go to completion. Estimates allow for the generation of 130 g endogenously through the processes of gluconeogenesis, that is from glycerol and amino acids. Hence, there is a shortfall of 50 g which would have to be provided exogenously in the form of dietary carbohydrate. In adult man, if total energy expenditure is 10 MJ and resting energy expenditure 7 MJ, the body requires the equivalent of 29% of total expenditure and 43% of resting expenditure in the form of glucose to prevent the development of ketosis (Table 6.1). The dietary intake of carbohydrate would have to be about 8% of total and 12% of resting energy expenditure. This calculation presumes that additional glucose can be generated from gluconeogenesis, which in the last analysis has to be derived from dietary protein, 15% of total energy, and the glycerol moiety of triglycerides. If the available protein or triglyceride were reduced then the exogenous demand would increase proportionately. However the sums are put together, the minimum requirement for

Table 6.1 A normal adult man leading a sedentary lifestyle requires about 10 MJ each day and has a resting energy expenditure of about 7 MJ, and 180 g carbohydrate is needed to limit ketosis (MacDonald, 1987).

Carbohydrate (g)		*Energy equivalent*		*Percentage*	
		kJ	*kcal*	*Total E*	*RMR**
Total needed	180	3010	720	29	43
Endogenous	130	2174	520	21	31
Exogenous	50	836	200	8	12

* Resting metabolic rate

carbohydrate represents a significant intake, and a substantial proportion of either total or resting energy expenditure.

So far as I am aware there are only a limited number of studies which can be used as a base of information to support these theoretical considerations. There is the need to explore the implications amongst a wider range of individuals and situations. I want to turn to a slightly different area which deals with a slightly different issue.

6.3.2 Energy and nutrients for normal growth

There is a problem in nutrition which is represented by a great tendency, an unreasonable enthusiasm, towards generalization and simplification. One area where this approach is seen most extensively, and takes on a most pernicious form, is the tendency to generalize observations made in adults to childhood. There is very little information of quality on the nutritional requirements for normal growth and development in children between the ages of 4 and 18 years of age. Even so, many groups have enunciated policy guidelines on diet and feeding for adult groups, and extrapolated these recommendations backwards from adults to children and adolescents. Even reports in which care has been taken to state categorically that the recommendations on healthy eating are only to be used in age groups above, say, 5 years, there has been a tendency to apply the recommendations willy nilly across the board to even younger children, leading in certain situations to inappropriate management and even pathological change. As a paediatrician I want to emphasize the danger of this casual approach. What do we know about the requirement for carbohydrate in the early growing period of life?

If we start at the beginning it might in fact be easier to identify absolute requirements for nutrients. Indeed, the standard approach to characterizing an adequate diet in adults is one which will sustain weight and nitrogen balance. In infancy and childhood there is the added condition that the diet must support normal growth and development. Expressed in other words, the diet must not only be able to maintain tissue mass and function, but must also be able to support net synthesis of tissue. In this context it is of some interest to note that glucose is not only the principal metabolic fuel for the developing embryo, fetus and newborn infants, but that in experimental animals fed on a low carbohydrate diet one can induce intrauterine growth retardation, a high incidence of still births and increased postnatal mortality. Against the background of the discussion presented by Professor Barker earlier, you should be able to appreciate the great caution which is required in promoting an intervention which might in any way interfere with fetal and infant growth. The potential for harm is not only limited to the short term, but also has to take longer-term considerations into account.

6.3.3 Evidence from animal studies

During 1986 there were reports which explored in some detail the nature of the relationship between dietary carbohydrate, dietary glucose and dietary gluconeogenic substrate on fetal growth and well-being (Koski and Hill, 1986; Koski *et al.*, 1986). These papers have not attracted a great deal of attention, but they are potentially of enormous significance and importance. It is true that the work has been carried out in experimental animals in the laboratory and there are all the problems of extrapolation between species, but these are, simply, not the sorts of studies which could be carried out in humans. Pregnant rats were offered diets which contained 0, 12 or 62% glucose and the effect on the outcome of the pregnancies explored. On the diet containing 0% glucose, mortality was as high as 40%. The paper also explored the difference in effect between giving a diet which contained triglycerides, and hence gluconeogenic precursors in the form of glycerol, and a diet in which fatty acids were the sole source of lipid, with no available gluconeogenic source other than the protein, which was 9.5% casein in all the diets. These studies showed that the maintenance of pregnancy required intact fat as 5 to 10% of the diet and carbohydrate as 4% of the diet, whether the carbohydrate was derived from glucose or as glycerol from lipid. Clearly, pregnant rats require sugars to carry pregnancy successfully. The sugars are essential at the cellular level. They may be provided preformed in the diet. If derived from endogenous production there is a dependence upon the availability of appropriate dietary precursors. Diets devoid of sugars or their precursors are associated with obvious, ultimate pathological changes.

It is of importance to know how subtle the responses to limited dietary sugars might be, and what is the lowest level of dietary and endogenous provision which can completely obviate any recognizable deleterious effect. Although, the maintenance of pregnancy required an absolute minimum of glucose at 4% of the diet, when rats were given 6 to 8% glucose in the diet they were able to sustain maternal food intake, maternal weight gain and normal fetal weight at term. If these three conditions are used as outcome indices, then 6 to 8% dietary glucose represents a sufficient minimal intake. If a more stringent outcome index were to be used, such as an adequate level of storage of carbohydrate reserves as glycogen, then the requirement is increased. In order to obtain fetal liver glycogen which is 50% of that seen in the offspring of mothers given standard chow during pregnancy, a dietary glucose of 12% is required. It is of interest to note that a similar proportion is needed in adults as the level needed for limiting ketosis.

For rats, lactation presents a greater demand upon metabolism than pregnancy as the newborn gains weight at such a rapid rate. In rats fed upon 0 to 6% glucose during late gestation, parturition and lactation, delivery might be delayed for days and mortality is increased so that by the fourth postnatal day 100% of pups are dead. Therefore, there are undoubted pathological consequences associated with low carbohydrate diets in situations of growth or where there is an increased metabolic demand. It is worth considering why this might be so, and possible mechanisms through which the pathology might be expressed.

6.4 METABOLIC DEMAND: SYNTHESIS

One of the problems which we face relates to some of the considerations which I raised earlier. In order to have a stable metabolic state it is necessary for intake to match demand. We have always restricted our concept of the demand for carbohydrates to a demand which will satisfy the requirements for energy. We have failed to consider, indeed almost virtually ignored, the need for dietary carbohydrate to act as the metabolic substrate for synthetic reactions associated with the turnover of metabolites of tissues and net deposition during growth (Fig. 6.3). It is reasonable to consider the nature of the synthetic reactions and the importance they may have for general metabolism, growth and development.

When you look in textbooks under carbohydrates to identify synthetic reactions, you will be fortunate if you find a section which deals with them specifically. More likely, you will have to hunt from page to page, putting together isolated pieces of information which are not presented with any coherence. This contrasts with the large sections which describe in great detail oxidation of glucose, elaborating at length upon the individual reactions and pathways associated with glycolysis, the tricarboxylic acid cycle and the hexose monophosphate shunt.

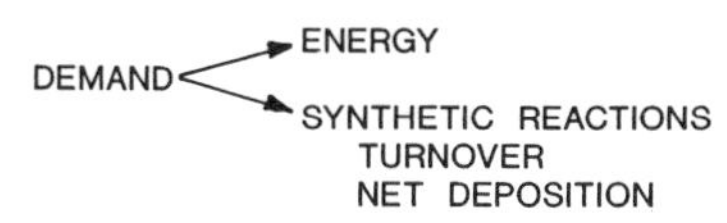

Figure 6.3 In a balanced metabolic state, when body weight and body composition are constant, intake must balance the demand. Demand includes the needs for energy as well as the requirements to satisfy synthetic reactions through the processes of turnover. During growth, demand also includes the need for net deposition.

Table 6.2 The metabolic demand for carbohydrates includes the requirements for the synthesis of metabolically active complex molecules, glycoproteins, glycolipids and proteoglycans

Glycoproteins:	most proteins having an extracellular exposure location function formed from complex chains of nine sugars involved in recognition, adhesion, receptors
Glycolipids:	sphingolipids and gangliosides specific receptor sites on cell membranes neuronal, synaptic transmission
Proteoglycans, mucopolysaccharides:	repeated disaccharide units 95% polysaccharide ground substance, intercellular cement cartilage, tendon, skin, synovial fluid

It is worth drawing attention to the synthetic reactions because it is likely that they are of substantially greater importance than we have credited in the past. Probably they represent one important variable expressing outcome which should be used when questions are asked about the need for dietary carbohydrate, either individually or collectively. There are three main categories of complex macromolecules which are synthesized from carbohydrates: glycoproteins, glycolipids and proteoglycans (Table 6.2). Glycolipids and glycoproteins are important components of cellular membrances (Fig. 6.4). The widely accepted perception of cellular membranes as a simple lipid bilayer, with a limited contribution from proteins which act as enzymes, receptors or carriers, represents only part of the truth.

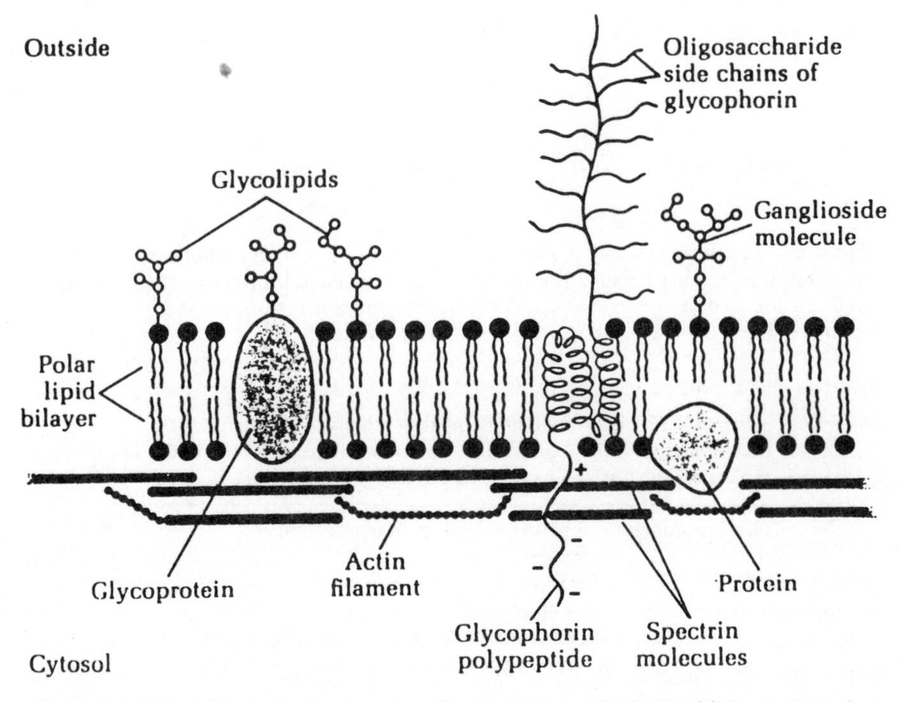

Figure 6.4 The cell membrane is a complex structure, which in addition to proteins and lipids contains compounds derived from carbohydrates, glycolipids and glycoproteins, which are essential for normal function (after Lehninger, 1982).

In the more complex models credit is given to the fact that both the lipid and the protein components of the membrane frequently contain side chains in the form of oligosaccharides. The side chains are critically important in terms of both structure and function. In reality, the membrane comprises a complex system of glycolipid, glycoprotein and lipoproteins. It is clear that some of the most important functional properties of membranes relate to the nature of the carbohydrate side chains.

Table 6.3 The oligosaccharides which comprise the membrane glycoproteins are formed from nine sugars

Hexoses	*Pentoses*	*Amino sugars*
glucose	arabinose	acetyl glucosamine
maltose	xylose	acetyl galactosamine
galactose		*N*-acetyl neuraminic acid
fructose		

6.4.1 Glycolipids

Glycolipids, in the form of sphingolipids and gangliosides, represent specific receptor sites on the surface of cell membranes and are also of critical importance for normal neuronal synaptic transmission.

6.4.2 Glycoproteins

The glycoproteins are, by and large, those proteins which have an extracellular exposure, location or function. Hence they are found in association with proteins on the outer surface of animal cells, or they are secreted to the cells' exterior or they are circulating proteins in blood or plasma.

Membrane glycoproteins are formed as complex oligosaccharides from nine sugars (Table 6.3). If you take a close look at these nine sugars, although some are readily recognizable, for the most part these sugars are not the sort of compounds which immediately jump into mind when considering dietetic factors or nutrition requirements for sugars. From a nutritional perspective this area of study is very poorly worked up.

Let me give an example. My attention was first drawn to the potential importance of these compounds a decade ago at the IUNS (International Union of Nutritional Sciences) meeting in San Diego in 1980. A good friend of mine, Dr Robin Whyte, asked me if I had had the opportunity to see the poster presentation on 'NANAs'. Bemused I thought he was referring to the diminutive of that most important Caribbean crop, bananas, and felt that national responsibilities require that I investigate further. In fact, he was referring to the work on *N*-acetylneuraminic acid being carried out by Dr Morgan's group exploring biochemical markers for functional and behavioural changes in malnutrition.

Morgan had suggested that the circulating levels of *N*-acetylneuraminic acid (NANA) might represent such a marker. NANA is a component of the gangliosides and glycoproteins which play important roles in neurotransmission, being integral components of the synaptic vesicles (Rahmann *et al.*, 1976). Hence these compounds are found in especially high concentrations in nervous tissue generally and nerve endings in particular; lower concentrations are found in plasma (Morgan and Winick, 1981). Rats given a low-protein diet demonstrate an abnormal pattern of behaviour and a reduction in the NANA content of the brain. There is a concomitant fall in the circulating concentrations of NANA, the magnitude of the fall correlating with the wet weight of the brain, ganglioside NANA and glycoprotein NANA concentrations in the brain. The administration of exogenous NANA to malnourished rats not only increases the concentration of ganglioside and glycoprotein NANA in the brain, but also causes a reduction in the associated behavioural abnormalities. If malnourished animals are exposed to a regimen of early stimulation, there is a reduction in behavioural abnormalities which is associated with an increase in both ganglioside and glycoprotein NANA concentrations in the brain.

Based upon these animal experiments, it was proposed that part of the developmental problems found in undernourished children might be associated with, or accounted for by, disorders of NANA metabolism. These ideas were potentially of great importance and very exciting. The possibility of solving the problem of delayed development in malnourished children by simple dietary supplementation was enormously attractive and of great social importance. The concentration of NANAs was measured in the blood of children at the time when they were malnourished, during recovery and again after recovery. The results appeared disappointing in that if anything the levels in the malnourished children were higher than after recovery, and there was certainly no rise with recovery, at a time when intellectual performance shows improvement (Hibbert *et al.*, 1985). These observations had the effect of reducing the level of enthusiasm for proposals which were underway for field studies involving supplementation of undernourished children with NANAs.

One conclusion which appears to have been drawn from a study in which the concentration of NANAs was measured in blood was that there was unlikely to be any benefit in supplementing malnourished children with NANAs. To my mind this is too simple a conclusion to draw. We know from amino acid studies that the concentration of a metabolite in blood does not bear any necessary relation to its turnover and metabolism. There is the need to explore in greater detail the possible effect of supplemental NANAs upon developmental performance in a controlled situation to identify whether they would have any effect. The metabolism of NANAs is complex and if one is to address the problem properly then there is the need to ask the right questions in the context of properly designed studies. It is naive to presume that you can enter into supplementation trials without having a clearer understanding of the metabolic interactions in humans. Certainly the rat studies imply that there is much more work to be done on our understanding of the importance of dietary sugars in metabolism.

Particular importance might be attached to the structure of the oligosaccharides because they have great potential for heterogeneity. Proteins are complex molecules formed from chains of amino acids, with the amino acids being joined through peptide linkages. The peptide linkage takes place through a single point in the amino acid molecule. Therefore, although there is quite considerable potential for variability in the pattern with which 22 amino acids may be joined together, the nature of the peptide bond does impose a limit to that variability. Peptides are essentially linear structures, which may be folded. There is little opportunity for the formation of branched compounds. In contrast, sugars can link together through a number of points in the molecule, thereby creating a wider range of potential linkages, and hence a richer variety of options. Even five sugars in a 13-residue oligosaccharide may form upto 10^{24} different polymers. The functional relevance of this is of enormous metabolic significance and underlines the important role played by membrane glycoproteins in the way the cell relates to its external environment, for example in cell recognition, cell adhesion, hormone function, receptor activity, nerve conduction and tissue typing.

Much of the present research focusing upon questions of cellular behaviour and intercellular reactions relates to considerations of the nature and function of membrane glycoproteins. The Science and Engineering Research Council (SERC) has supported a major research activity which explores the nature of membrane function, and yet nowhere within this wide canvas of understanding is there any consideration given to the question as to whether dietary intake of individual sugars or any particular combination of sugars exerts any modulating influence upon the behaviour of cell membranes. We have little information on the relationship between the dietary intake, nutritional significance and metabolic handling of most of the nine essential sugars identified as fundamental components of oligosaccharides. This is an area of enquiry which we have ignored to our cost. It is an area of understanding which could bring us into the mainstream of the cutting edge of scientific enquiry. We cannot afford either to ignore it or to walk away from it. At the present time we have not even bothered to recognize its existence.

6.4.3 Mucopolysaccharides and proteoglycans

The third and final group of complex sugars is the mucopolysaccharides or proteoglycans. These compounds are formed from a protein core with peptide side chains and the side chains have associated polysaccharides with them. The polysaccharides in this formation have a number of important chemical and physical properties, amongst which the ability to hold or to bind water might be of special importance. This varied group of complex materials comprises a range of the most important extracellular compounds in the body. Disorders of mucopolysaccharides and proteoglycans have been associated with a number of disease states and pathological conditions.

The more complex molecules are derived from repeating disaccharide units and although they are built on a protein core, about 95% of the final molecule consists of polysaccharide. Basement membranes, ground substance and intercellular cement are all made from this particular group of compounds. They also form an integral part of cartilage, tendon, skin and synovial fluid, as well as being the main constituent of hyaluronic acid and chondroitin sulfate, etc. Again if we ask the simple question as to how their production, turnover and degradation might be modified either by the dietary intake or the nutritional state we are met with a numbing silence, representative of total ignorance.

Pregnant women, infants, children and adolescents all need to grow and to be capable of the net deposition of complex carbohydrates, of importance for structure and function. We have no knowledge about how dietary sugar intake might influence these processes.

6.4.4 Glycogen

The one synthetic activity which dietary sugars are generally seen as participating in is the formation of glycogen, and the important role it plays as a source of glucose in the short term when the demand for glucose exceeds the supply.

A large part of the popular concern about the potential damaging effects of carbohydrates and sugars in the diet is driven by a disease-oriented perspective generated by practising clinicians. Doctors treat sick patients and most of the people who doctors see are sick. Care has to be taken in extrapolating from observations made in this group of individuals to the rest of the population at large. In general, most of the people in the world are healthy and seldom present to their doctor, if at all. We have only a very limited understanding of the sugar or carbohydrate requirements for ordinary healthy people, because we have never bothered to ask the question in that way. Let me illustrate with an example.

We have already identified that a normal adult male with a relatively sedentary life style might ingest and expend in the order of 10 MJ per day. If this is taken as a mixed UK diet then the likelihood is that the individual will satisfy his requirement for all other nutrients, proteins, vitamins and minerals. Were this individual then to engage in an increased level of activity, say to expend an additional 4 MJ per day, we are advised that there is no need for additional protein, minerals or vitamins to go with the increased ingestion and expenditure of energy. The question is therefore: 'Does it matter in what form the additional 4 MJ per day is ingested?'. So far as I am aware we do not have a complete answer to the question, but there are maybe two statements which one can make with a degree of confidence. First, if a substantial proportion of the extra calories is not provided in the form of carbohydrate, then the individual will not be able to sustain the same level of performance over any extended period of time. Sustained performance requires that during recovery from exercise the glycogen stores in muscle must be repleted, and this requires dietary carbohydrate. Second, we do not know whether it makes any difference if the additional carbohydrate is provided by simple sugars or by more complex carbohydrates. At last I reach the point, which I presume this debate is all about.

6.5 DIETARY CARBOHYDRATE

The question about the relationship between simple sugars and complex carbohydrates, is, I think, what Dr Eastwood really wanted me to discuss. I have taken a long time and a circuitous route to get there. This most difficult question clearly represents one part of the nub of the problem. My greatest difficulty is that I do not think we have adequate answers to some of the most simple and straightforward questions. My hope is that 'Nutrition in the Nineties' will develop a more scientific approach to the way in which questions are asked, so that in 10 years' time the debate can take place from a more informed, less opinionated, base of knowledge and understanding. In the simplistic view of carbohydrates, they may be divided into four categories, although this classification may not be sufficiently detailed to satisfy the cognoscenti.

1. There are non-starch polysaccharides (NSP), which may or may not be fermentable in the large bowel by the colonic microflora. The non-fermentable NSP passes through the bowel unchanged. Fermentable NSP is acted upon by the colonic microflora and has a complex, as yet not fully characterized fate, which includes the production of short-chain fatty acids which can be used by the host.
2. There are non-digestible starches, which because of their particular structure and physico-chemical properties at the time of ingestion resist digestion in the small bowel, yet are likely to be fermented by the microflora in the colon with a fate similar to that of fermentable NSP.
3. There are digestible starches which are subject to digestion by secreted enzymes from the oral cavity down to, and including the colon. The pattern of digestion may vary, depending among other things on the extent to which the food is chewed and the extent to which salivary and small intestinal amylase is allowed to act. The products of the digestion of starch are absorbed as simple sugars.
4. There are simple sugars, which can be absorbed either unchanged or with only minimal digestion.

One point upon which the literature is far from clear is the physiological importance of salitvary amylase and the time over which it is allowed to act. The extent of the activity of salivary amylase would seem to be one important variable in those studies in which the glycaemic index of different sources of carbohydrates or different diets is assessed. To what extent is the rate of extent of the digestion of starch significantly altered in individuals who bolt their food when compared with those who allow a reasonable time for digestion to take place in the oral cavity and probably for a limited time in the stomach? There is no good evidence to suggest that the utilization of simple sugars produced by digestion of complex carbohydrates is different in any way from simple sugars which are ingested as such. Are the perceived risks to health which have been associated with the ingestion of simple sugars in fact a spectrum of risk with simple sugars at one extreme, moving through digestible starch to fermentable starch and fermentable NSP to non-fermentable NSP at the other extreme? If so, what is the detailed basis of this risk?

It is generally assumed for humans that short-chain fatty acids are unlikely to be effective sources of gluconeogenic precursors, and therefore one cannot assume that fermentable starch or fermentable NSP can function as a metabolic alternative to simple sugars, digestible starch or other gluconeogenic substrates. However, there is evidence from both ruminants and horses that propionic acid may be gluconeogenic, making a significant contribution to endogenous glucose production (Ford and Simmons, 1985). What happens to individuals on a 'high fibre diet'? Are they at risk of running short of glucose units for cellular metabolism? Is it true that the only difference between simple sugars and digestible starches is a consideration of the rate, nature and site of digestion of the starch? If the glucose or sugars for metabolic interaction has to derive from one or other of these dietary forms it may be that the only important difference between them is the extent to which an individual does or does not chew their food.

6.6 CONCLUSION

Our approach to the understanding of dietary sugars and other carbohydrates has fallen into the trap of reducing our perception of the world to simplicity in order to be in a position to make sweeping generalizations which we perceive as being valuable to policy-makers. There is the need to adopt a nutritional, rather than a dietetic perspective, and to ask questions of metabolism within the context of the basic principles of nutrition. We need to be able to define the extent to which carbohydrates in general, or individual carbohydrates in particular, may be absolutely or conditionally essential. We need more information on the minimal requirement for carbohydrates as an important source of energy for the body in general, and how this might be altered by the availability of alternative sources of energy. In this context we need a finer appreciation of the turnover of carbohydrates at both the whole body and tissue levels. We have very little understanding of the extent to which changes in the pattern of dietary carbohydrate might influence membrane and cellular function. There is a need to build linkages between the basic scientific enquiry in these areas and human nutrition.

Present day understanding in the wider biological sciences offers many rich opportunities for enquiry. If taken, these opportunities offer the possibility of bringing nutrition into the mainstream of scientific thought in the 1990s. The approach cannot simply be based upon phenomenological observations, but requires a rational investment in basic studies to explore complex metabolic inter-relations at the level of the whole body. This sort of work costs money and takes patience, time and effort. The 1980s showed that money seldom followed patience, time and effort. Let us hope that the 1990s are more discerning in the values that society adopts.

REFERENCES

Committee on Medical Aspects of Food Policy (COMA) (1989) *Dietary Sugars and Human Disease*, Report on Health and Social Subjects No. 37, HM Stationery Office, London.

Ford, E.J.H. and Simmons, H.A. (1985) Gluconeogenesis from caecal propionate in the horse. *Br. J. Nutr.*, **53**, 55–60.

Gregory, J., Foster, K., Tyler, H. and Wiseman, M. (1990) *The Dietary and Nutritional Survey of British Adults*, HM Stationery Office, London.

Hibbert, J.M., Jackson, A.A. and Grantham-McGregor, S.M. (1985) Plasma concentrations of *N*-acetylneuraminic acid in severe malnutrition. *Br. J. Nutr.*, **53**, 11–16.

Koski, K.G., and Hill, F.C. (1986) Effect of low carbohydrate diets during pregnancy on parturition and post natal survival of the newborn rat pup. *J. Nutr.*, **116**, 1938–48.

Koski, K.G., Hill, F.W. and Hurley, L.S. (1986) Effect of low carbohydrate diets during pregnancy on embryogenesis and fetal growth and development in rats. *J. Nutr.*, **116**, 1922–86.

Lehninger, A.L. (1982) *Principles of Biochemistry*, Worth, New York.

MacDonald, I. (1987) Metabolic requirements for dietary carbohydrate. *Am. J. Clin. Nutr.*, **45**, 1193–6.

Morgan, B.L.G. and Winick, M. (1981) The subcellular localization of administered *N*-acetylneuraminic acid in the brains of well nourished and protein restricted rats. *Br. J. Nutr.*, **46**, 231–7.

Rahmann, H., Rossner, H. and Breer, H. (1976) A functional model of sialo-glyco-macromolecules in synaptic transmission and memory formation. *J. Theoret. Biol.*, **57**, 231–7.

DISCUSSION

James On a question of essential and non-essential amino acids, it has been stated that the non-essential ones must be somewhat essential as the body makes them from the other group.

Mela, Reading I have a number of comments I would like to make to Dr Heaton. The first is regarding his concept of the aetiology or speculative aetiology of Crohn's disease. It is possible that, in fact, he has causation reversed, i.e. that patients may change their dietary habits or sensory preference secondarily to having Crohn's disease. There are certainly examples of patients, such as those with Addison's disease, who increase their salt preference and consumption of salt. Subjects with bile insufficiency spontaneously decrease fat intake.

The main issue that I wanted to comment on was that of caloric intake and the effect of sugar or reduction of sucrose intake and hence caloric intake. In the Puerto Rico study the subjects were consuming over 4000 calories during the period before they went on the aspartame regime. They also were given no choice but to consume the aspartame-containing products. There are much better population and better clinical data which suggest that people will indeed compensate for such changes. Since all carbohydrates, including simple sugars, are absorbed as their constituent monosaccharides, his hypothesis presumes that the difference in energy intake between simple and, say, complex carbohydrates is due to sensory preferences overriding metabolic signals. The sensory literature, I would say, does not support that notion. Obese subjects seemed to show a decreased preference for sweets and no difference in sugar intake. Thirdly, the biochemical evidence does not support this. For one thing there seems to be a limited ability to convert carbohydrate to fat in humans and this process is only 75% efficient. Thus certainly sucrose or sugars relative to fat are going to be less likely to be a source of body fat. Lastly when the subjects are placed on a low sugar diet I am not sure what foods they are avoiding, but many of the common sources of sweetness and sugar are in fact rich in fat. Certainly that is true of pastries, cakes, cookies, chocolate, confectionery.

Heaton The idea that the higher sugar intakes of patients with Crohn's disease are secondary to the disease has been looked at in a variety of ways and there is nothing to support it. In particular, people have asked patients not only about their current intake but also about their lifetime or previous intake. The trend in four out of five studies was towards a higher sugar intake before they had their disease, before they had any symptoms. This suggests that having Crohn's disease tends to decrease sugar intake, not increase consumption.

You say population data shows aspartame is compensated for. I do not understand how population data can be used to show this. Surely you have to have an experiment if you want to look for compensation.

You say obese people have a decreased preference for sweet foods. I do not think that is relevant because obese people are people who for years have been suffering severe self-image problems and social problems from being overweight and have made every kind of effort to do something about it and have distorted their eating habits in reaction to their obesity. So their current eating habits and preferences tell us nothing about what made them obese in the first instance.

You say sugars are less efficient than fats at being turned into fat. That is a biochemical fact of course but it does not address the issue that we are discussing – whether extracellular sugars do make you consume more than your required physiological number of calories. That is a totally separate issue. Finally, I agree with you that many sugar-rich foods are high in fat. I see that not as an argument against sugars being important in calories imbalance, but rather an argument for saying that they are even more important because so often when people go for something sweet, unconsciously or not, they take in fat as well.

Powell-Tuck, London There is a danger that we will overplay the Crohn's disease story, but since it has been mentioned twice can I just support the last questioner. With respect, I am afraid I disagree with Dr Heaton. I think all the evidence is that Crohn's disease patients are in fact self-selecting a low fibre diet. The one prospective study that we have from Jean Ritchie, which of course was multicentered and came out of Ken Heaton's excellent stimulus in this direction, has shown above all things that patients with Crohn's disease, if put on a high-fibre diet, get abdominal pain and bloating. I think the one thing that we do learn from the Crohn's disease studies is that patients try to avoid fibre. I agree with the previous questioner that the most likely explanation for the high refined sugar intake in Crohn's disease is self-selection. It is a result of the disease rather than the cause. It is always interesting to hear the other possibility though.

Heaton I have yet to see any data other than the gut feelings you are expressing which support the idea that patients have reacted to their disease by increasing their sugar intake. Jean Ritchie's admirable study was a therapeutic trial. It was nothing to do with spontaneous sugar intake.

Eastwood Professor Jackson, could you expand on the metabolic consequences of taking glucose either through sucrose or through starch, two contrasting forms, one a slow release form and the other a readily absorbed form. The metabolic response and perhaps a whole sequence of events after that will be affected by the availability, surely this is one of the major points in this sugar versus other sources of carbohydrate dilemma.

Jackson I think that a large number of the studies in which comparisons have been made are difficult to interpret simply because one of the important variables to which I have alluded is that the extent to which salivary amylase is allowed to act has not been standardized. Unless you control for that variable, I do not see that you can reasonably interpret your results. Undoubtedly, if you provide a metabolite along a metabolic chain you go from a complex polysaccharide, starch, or an oligosaccharide down to a monosaccharide or a disaccharide. The farther you are down that process when the food is ingested, the less time it is going to take. Now the question of the rate of the metabolic response and the nature of that is then determined by a number of factors other than how far down the metabolic chain you are, and not least is the form of the diet in which that compound is taken. I think it seems quite clear that there are substantial differences depending on the dietary form, and I am sure Ken would argue that if you take sucrose as sucrose without any other dietary component then you get different effects. He would argue that there are potentially adverse effects. I think that there are potentially adverse effects, but whether or not they are genuinely adverse depends upon the level of sugar, not only within a single meal but also over a day against the background of the total diet. The margin for normal response is substantial although there is an upper limit to that and it can be exceeded. But for the majority of people it is unusual to exceed that upper limit.

Heaton I think there are two quite separate questions; one is 'do people react differently to sugars as opposed to starches?' and the other is 'do they react differently to sugars in extracellular form as compared with sugars in their natural or intracellular form?'. I think there is fairly good evidence that there are differences in both regards. The sugar versus starch issue I regard as less important because the experiments that have been done, although very good, have been limited to a subsection of the population. I am referring, of course, to the extensive work of Sheldon Reiser and his colleagues at Baltimore who showed that in this subsection about 10 to 15% of the population selected on the basis of a rather brisker than average insulin response to a sugary challenge, if they are fed large amounts of sugar as opposed to starch for several weeks they develop hyperlipidaemia, sometimes hypercholesterolaemia, a fall in HDL cholesterol and evidence of insulin resistance with impaired glucose tolerance. In fact most of the features of what Gerald Reaven has called syndrome X and which he sees together with truncal obesity as an important risk factor for coronary heart disease. These people clearly pay metabolic penalties for high sugar intake. Taking sugars in an intracellular rather than extracellular form has benefits not just in a minority but probably in the whole population. In normal subjects, we have shown repeatedly that meals based on intracellular sugars induce far less insulin than meals based on extracellular sugars with exactly equivalent amounts of sugars.

Wharton, Glasgow In the 1970s there was concern about the fructose in sucrose and its effect on lipid metabolism, and cholesterol, I think it was referred to in the COMA report. Is all that concept discounted?

Heaton I do not think it is discounted. I think fashion has just moved on to another area of interest, that is all.

Wharton It is just taken as read?

Jackson Certainly, there are considerable biochemical differences between fructose and glucose in experimental animals, between sucrose and starch on one hand and glucose and fructose on the other hand.

Macdonald Fructose is handled differently compared to glucose and fructose whether it is in sucrose or not, but it seems that the body adapts to it, except in people with hypertriglyceridaemia.

Wharton So if you are bothered about the effect of fructose on lipid metabolism, then to replace sucrose just by glucose has no particular advantage.

MacDonald Except for those people who have hypertriglyceridaemia, yes they should perhaps avoid sucrose or fructose.

Cannon, London I would like to support Dr Heaton in a remark he has just made about food (as distinct from nutrients), concerning sugar and fat. Not long ago, the President of the Biscuit, Cake, Chocolate and Confectionery Alliance boasted to the Minister of Agriculture and Food that the products manufactured by his Alliance account for one-eighth of all consumer expenditure on food and drink in Britain. To a considerable extent, therefore, it is artificial to talk about the public health problems of sugars separately from those of fats. Sugar makes fat, especially saturated fats, palatable.

Winkler, London Professor Jackson, repeatedly throughout your talk there was an elusion between carbohydrates, glucose and sugar and you were making the case about the essential requirement for glucose. Can you clarify? Are you suggesting that there is an essential requirement for simple sugars?

Jackson I spent some time in trying to be cautious about how the word essential is used and quite clearly at the level of the cell, at least for some cells, if not for all cells, there is an essential requirement for simple sugars. How the body seeks to provide the cells with those simple sugars is varied and depends upon the diet and does not of itself require that simple sugars are present in the diet.

Moynahan, London Have you found any evidence from the patients you have studied that the HLA type can be associated with increased consumption of sugar?

Jackson I am not in a position to answer that question, I do not know whether anybody in the audience has a suitable answer.

Nicol, Reading I would like to make one or two remarks about Dr Heaton's paper. Is it possible that the caloric intake reduction when sugar was taken out of the diet was due to unpalatability because sugar is, as he mentioned, a very good palatability agent. I would like to comment on the feasibility of taking sugar as sugar cane, as once sugar cane is cut, microbiological degradation rapidly sets in. Next he mentioned that sugar in drinks could be of the order 20 g per 100 ml. I have yet to find anyone who has tea and coffee at such a high level of sweetness. The normal level of sweetness in tea and coffee is about 5 or 10 g per 100 ml.

Heaton The Italian coffee cup is about 30 or 40 ml and they put two or three teaspoonfuls of sugar into it! The point I was really making was that it is the amount of sugar you take that matters, not necessarily whether it is in a drink. Half a Mars bar would give you 20 g of sugar and many people would take that dry without a drink. The point is really the gastric osmolality, not that of the drink. Decreased energy intake, due to decreased palatability? This obviously must be a factor. One of the reasons sugar makes us overconsume calories is because we like it so much. All the same, a distinction must be made between a ripe Williams pear or some mango or lychees, where all the sugars are intracellular, and delicious Scottish cakes.

PART THREE

Fibre

7 Towards a recommended intake of dietary fibre

J.H. Cummings and S.A. Bingham

7.1 INTRODUCTION

It is our brief to make the case for fibre. This we will try to do and we will even suggest a recommended intake for dietary fibre for the community.

The list below shows the now familiar disorders allegedly due to deficiency of fibre in the diet:

- coronary heart disease
- diabetes mellitus
- obesity
- gallstones
- hiatus hernia
- varicose veins
- large bowel disorders
  - constipation
  - irritable bowel disease
  - colonic diverticulosis
  - colonic polyps and cancer
  - appendicitis
  - haemorrhoids

The concept of fibre-deficiency disorders was popularized by Burkitt, Trowell and others about 20 years ago (Burkitt, 1969; Trowell, 1972). The idea was immediately attractive because they were offering new ideas about very common diseases, for many of which there were no really plausible explanations as to their aetiology. It was for this reason, rather than the appearance of a major body of experimental evidence, that the idea rapidly became disseminated.

Initially there was much interest and research into dietary fibre, so that by 1980 the Royal College of Physicians published a report entitled *Medical Aspects of Dietary Fibre* (Royal College of Physicians, 1980) which concluded

'On present evidence, we think it highly probable, though not fully proved and possibly not susceptible of rigid proof, that increasing the proportion of "dietary fibre" in the diet of Western countries would be nutritionally desirable'.

Since 1980, enthusiasm for fibre has waned somewhat for a variety of reasons. The initial fervour with which the fibre story was propounded, and its uncritical acceptance, failed to convince many people in the scientific community that this was a genuine hypothesis. Furthermore, the public rapidly caught on to the concept and have constantly pressed for recommendations to be made about fibre in the diet. This has led to often poorly substantiated claims for high-fibre foods. A more real problem has been the great difficulty of doing research into dietary fibre since it comprises a complex mixture of plant polysaccharides with both physical and chemical properties. It is particularly the understanding of the physical properties of fibre that has led to so many errors in research studies and a general failure to unravel its physiological effects.

An equally important problem in the progress of the dietary fibre story has been the lack of an internationally agreed definition of fibre. A number of very different methods have arisen for its measurement in food, but they give quite varying results in any single commodity and thus serve to increase the confusion. At the present time this is a major impediment to progress and is unlikely to be resolved easily as national and commercial interests compete for what they see as a highly saleable dietary fraction. A logical way out of this impasse has recently been recommended in the report of the British Nutrition Foundation Task Force (1990) on complex carbohydrates in foods. In this the '... Task Force recommends that the word 'fibre' should become obsolete at least in the scientific literature'. The principal reason for this conclusion in the Task Force report is to encourage scientists to define accurately the physical and chemical properties of the various complex carbohydrates in the diet and then look specifically at their biological effects. This recommendation takes into account the recent interest in resistant starch which complicates the chemical measurement of dietary fibre, but which physiologically overlaps to some extent with the effect of certain cell-wall polysaccharide fractions. In the future, therefore, it is hoped that there will be fewer papers recording the effects of 'dietary fibre', but more looking at starch, resistant starch (RS) and non-starch polysaccharides (NSP) and at their physical and chemical interactions in the body.

7.2 THE CASE FOR DIETARY FIBRE (NON-STARCH POLYSACCHARIDES) IN THE DIET

Dietary fibre has been implicated in the aetiology of a number of diseases. Let us examine the need for its inclusion in the British diet.

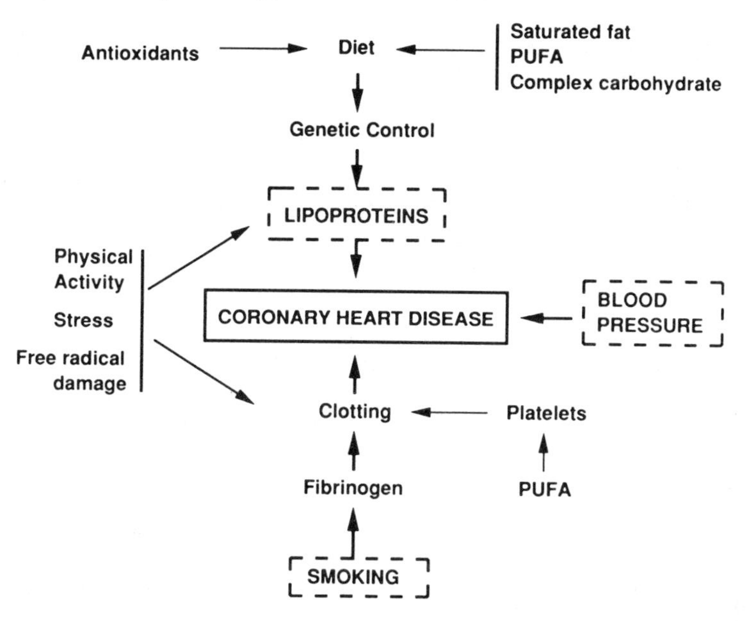

Figure 7.1 Major risk factors and other contributory causes to coronary heart disease. PUFA = polyunsaturated fatty acids.

7.2.1 Coronary heart disease

Coronary heart disease is a condition in which there are a variety of cellular and pathological events leading ultimately to a coronary occlusion and often to death. The major risk factors are a raised blood pressure, raised serum cholesterol and smoking. A number of other processes also relate to risk in this condition and these are summarized in Fig. 7.1.

From this figure it can readily be seen that diet itself is only one of a number of causes contributing to risk of coronary heart disease, and furthermore that fibre is only one of the many dietary components which affect risk. What, therefore, is the evidence that fibre may be a significant dietary influence in this condition?

(a) Physiology

One important credential is the now well-established ability of certain forms of NSP to lower blood cholesterol in man. These studies have been reviewed extensively elsewhere (Anderson *et al.*, 1990) and in summary show that soluble forms of NSP such as pectin and guar gum and the soluble β-glucans in oats will consistently lower blood cholesterol in healthy and hyperlipidaemic subjects. Insoluble forms of NSP, such as that present in wheat bran or pure cellulose, do not have this effect. The possible mechanisms whereby NSP may lower blood cholesterol are several and not yet conclusively established. They include an effect of the water-soluble gel-forming materials in inhibiting cholesterol absorption in the small intestine, or affecting bile acid absorption and metabolism. Also, soluble NSP is readily fermented in the large intestine producing short-chain fatty acids, including propionic acid. There are some *in vitro* and animal studies to show that propionic acid may be hypocholesterolaemic. Finally, soluble NSP has notable effects on blood glucose and insulin metabolism and through this may also indirectly affect lipoprotein metabolism.

(b) *Epidemiology*

In a number of epidemiological studies attempts have been made to correlate fibre intake with heart disease risk. Temporal correlation studies have not provided convincing evidence for a protective effect of fibre, but these studies can be criticised for the nature of the fibre intake data that are used. Mostly only very imprecise intake measurements have been made and there are really no good cross-sectional studies using accurate measurements of NSP intake.

A number of prospective studies have been done of diet and heart disease and these show a much more consistently protective effect of fibre. These include the studies by Morris *et al.* (1977) in London, Yano *et al.* (1978) in Honolulu, the Zutphen study (Kromhout *et al.*, 1982) and the Boston-Irish study (Kushi *et al.*, 1985). In all these, some measure of total fibre intake has been made and the coronary heart disease cases have had uniformly lower intakes. However, the results are confounded by the close inter-association between dietary patterns which reflect high fat, low carbohydrate and vice versa. There is a clear need for more work to be done using good methods of dietary intake assessment and precise measurements of all the carbohydrates in the diet.

Overall, therefore, with regard to fibre and coronary heart disease we have a multifactorial problem for which there is some evidence for a contribution from fibre. However, on the basis of these studies alone, there is probably not sufficient evidence to make a specific recommendation to the population for an intake of fibre. What can probably be said with more certainty is that diets characterized by high starch, NSP and low fat will be beneficial.

The same general arguments can probably be applied in relation to fibre and diabetes, obesity, gallstones, etc. Fibre may be a contributory component to their aetiology, but it is difficult to make a case for an essential role in the prevention of these conditions on present evidence.

7.2.2 Large bowel disorders

When it comes to large bowel disorders it is clear that fibre has a greater role to play. The actions of NSP on the large intestine are well established and are listed below:

- increases stool weight;
- shortens intestinal transit;
- stimulates microbial growth;
- decreases ammonia and phenol levels;
- provides short chain fatty acids;
- alters bile acid metabolism;
- influences mucosal growth.

Non-starch polysaccharide together with resistant starch is probably the most important controlling factor for large bowel function. This is in contrast to small bowel function where so many dietary components contribute to the control of carbohydrate and lipid metabolism. It is therefore much more likely that if fibre is going to have a major impact on health it will do so through its effect on large bowel diseases.

What are the risk factors for large bowel disease and how can we judge the need of a population for NSP in the diet? For coronary heart disease the major risk factors are known and readily measured, but when it comes to the large bowel there are no well-established criteria. The list below gives an indication of some of the measurements which might be made at a population level to identify large bowel risk.

- bowel habit
- transit time
- breath hydrogen and methane
- genetic analysis
- dietary intake
- invasive techniques
    - X-ray
    - microcapsules
    - histology
- urinalysis
- faecal analysis
    - short-chain fatty acids
    - occult blood
    - mutagens
    - ammonia, phenols, amines, sulphide
    - bacteriology
    - bile acids

Of the various risk factors, bowel habit and in particular daily stool weight is probably the best global external reflection of events within the large bowel. There are also a lot of data about bowel habit and the various influences on it, although it is much more difficult to get an accurate measurement of somebody's average daily stool weight than is often acknowledged.

7.3 BOWEL HABIT AND BOWEL DISEASE RISK

7.3.1 Constipation

Let us therefore look at bowel disease risk in relation to bowel habit, starting with the obvious problem of constipation. Constipation is a disorder of bowel motility characterized by the passage of small amounts of hard stool infrequently and with difficulty. It is a common problem in the UK, but not one that attracts large grants from funding agencies. Figure 7.2 shows consultation rates for constipation in general practice in the UK for the years 1981–2 (Royal College of General Practitioners, OPCS/DHSS, 1986). The figure shows that constipation is common in infancy and in old age, with higher rates in women than men during middle life. Overall these data accumulate to about half a million people consulting their GP each year on account of this problem. In population surveys the prevalence of constipation is even higher.

In both the UK and USA, 1% of the population consult their family doctor

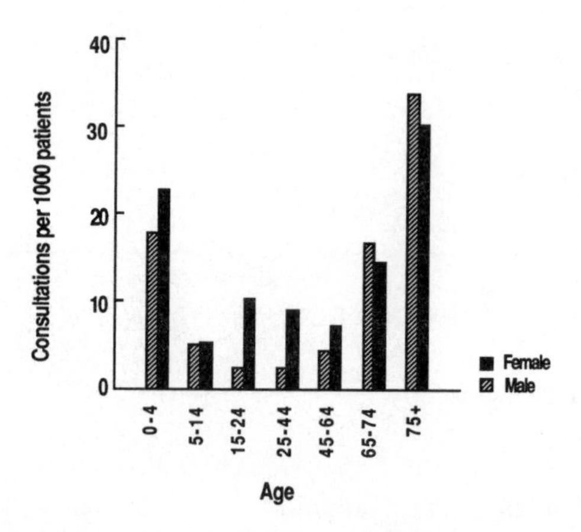

Figure 7.2 Number of patients (per thousand) consulting their family doctor in the UK for constipation (from Royal College of General Practitioners, OPCS/DHSS, 1986).

annually because of constipation and in surveys 10–12% consider themselves subject to symptoms of constipation (Royal College of General Practitioners, OPCS/DHSS, 1986; Sonnenberg and Koch 1989; Sandler *et al.*, 1990). The prevalence is much greater in people over 60, with 20–30% constipated and taking laxatives from time to time (Connell *et al.*, 1965; Thompson and Heaton, 1980). In Connell's study, 4% of an industrial population of 1055 persons thought themselves to be constipated and 16% of 400 people in general practice did. Six per cent of the total study group took laxatives weekly and nearly 20% occasionally. In the older age group laxative consumption was 30%. In Thompson and Heaton's study of 301 healthy volunteers, 3% of the young, 8% of the middle-aged and 20% of the elderly complained of constipation. Even higher proportions were found in a study of 350 outpatients at an ear, nose and throat hospital (Moore-Gillon, 1984).

There are of course problems in defining and measuring constipation and in particular stool output is rarely measured. However, Fig. 7.3 shows the results of bowel habit data in eight published studies. Average stool weight for the constipated people is about 50 g/day. These data clearly overlap considerably with those of people in the UK who have less than average daily stool weight. Cowgill and Anderson's study in 1932 is particularly interesting. They took healthy men and measured their bowel habit whilst on a variety of NSP intakes. When the subjects were given bran their stool weights were 193 g/day and they declared that bowel habit was satisfactory. However, when NSP intake was 10 g/day stool weight was 114 g/day and bowel habit was

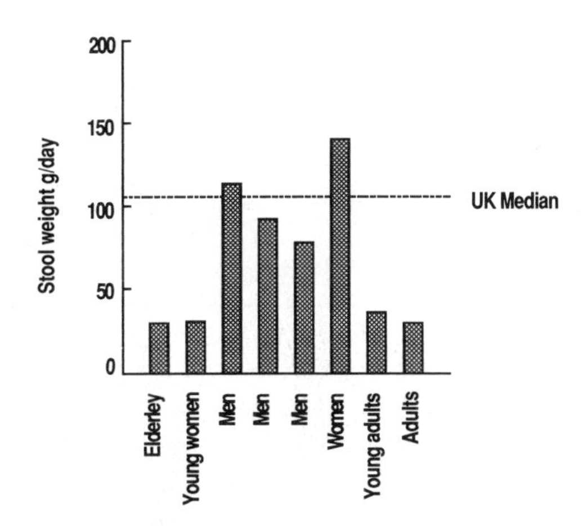

Figure 7.3 Average daily stool weight (with SEM) in eight groups of people complaining of constipation. From Cowgill and Anderson (1932); Cowgill and Sullivan (1933); Smith *et al.* (1980); Bass and Dennis (1981); Meyer and Le Quintrec (1981); Graham *et al.* (1982); Marlett *et al.* (1987). Median stool weight for the UK population is also shown from Cummings *et al.* (unpublished data).

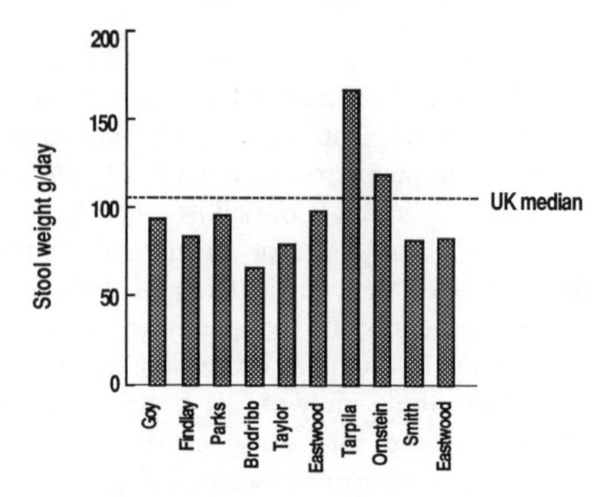

Figure 7.4 Average daily stool weight in 10 groups of people with symptomatic diverticular disease. From Goy *et al.* (1976); Findlay *et al.* (1974); Parks (1974); Brodribb and Humphreys (1976); Taylor and Duthie (1976); Eastwood *et al.* (1978a); Tarpila *et al.* (1978); Ornstein *et al.* (1981); Smith *et al.* (1981); Eastwood *et al.* (1978b). Median stool weight for the UK population is also shown from Cummings *et al.* (1991).

described as 'unsatisfactory' and 'mild constipation'. At an NSP intake of 6 g/day stool weight was 93 g/day and bowel habit described as 'very unsatisfactory'.

Marcus and Heaton (1987) have shown that when constipation is induced experimentally irritable bowel-like symptoms occur. Furthermore, if healthy subjects are put on NSP-free diets then constipation is a common complaint and stool weight is usually in the range of 40–60 g/day. Overall therefore constipation becomes an increasingly common problem at stool weights below 100 g/day and some subjects even consider their bowel habit unsatisfactory when it is in the range 100–140 g/day.

7.3.2 Diverticular disease

Diverticular disease of the colon is also common in the UK and its aetiology is thought to be very much associated with low stool weight (Painter, 1975). In the study of Gear *et al.* (1979) in a symptom-free population in Oxfordshire studied by X-ray examination of the colon, 12% of those aged 45–49 years had diverticula and in the 70+ age group 64% were affected. In the Third World, diverticula are rarely found in those centres where autopsies are performed (Segal *et al.*, 1977). These and other epidemiological data suggest this is an acquired disease of the Western world.

In a number of studies bowel habit has been measured in diverticular

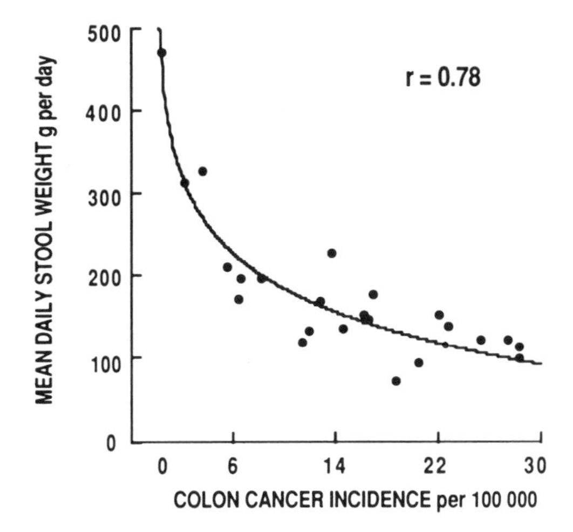

Figure 7.5 Mean daily stool weight for 23 populations from 12 countries in relation to colon cancer risk (age standardized to a world population). From Cummings *et al.* (1991).

disease. Figure 7.4 shows stool weights from 10 such reports. The study of Tarpila stands out as being somewhat different from the rest and by excluding it the average stool weight in this population is about 90 g/day (n = 275). These studies often included both symptomatic and asymptomatic patients, some of whom reported suffering from diarrhoea. Overall it looks as though patients with this condition, or in the process of developing it but being asymptomatic, have stool weights less than 100 g/day.

7.3.3 Large bowel cancer

A third disease in which fibre and bowel habit may play a role is bowel cancer. Burkitt popularized the view that bowel cancer was associated with low stool weight in a paper in 1971. He measured stool output and transit time in many population groups and related these to national mortality and morbidity statistics, where available. In general, low stool weight and slow transit time were associated with greater risk of bowel cancer. The proposals aroused much controversy and debate, largely because the measurements made did not entirely stand up to the exacting criteria required of modern epidemiology. They nevertheless have provided the impetus for a number of subsequent studies.

In a report of bowel habit and bowel cancer risk in Scandinavia (Cummings *et al.*, 1982) low stool weights were shown to be associated with an increase in bowel cancer risk. In a more recent and comprehensive analysis Cummings *et al.* (unpublished data) have shown that there is an inverse relationship between stool weight and bowel cancer risk when populations from 12 or more countries, for which bowel habit data are available, are examined.

A further dimension to this story relates to the effect of constipation on bile acid metabolism. High levels of secondary bile acids were thought at one stage to be a risk factor for bowel cancer (Hill *et al.*, 1971, 1975). Whilst some support from animal experiments was provided for this theory, more recent epidemiological data have not entirely supported the hypothesis (Setchell *et al.*, 1987). Nevertheless, if constipated subjects are given laxatives to speed up transit time and increase stool weight the amount of deoxycholic acid in bile falls (26% ± 8.6% constipated to 17% ± 8% with laxative) (Marcus and Heaton, 1986). Furthermore, in another study of Marcus and Heaton (1987), when healthy subjects had symptomatic constipation induced and transit time increased from 48 to 103 hours the deoxycholic acid pool size increased significantly.

From the bowel cancer data it looks as though bowel cancer risk is high when stool weight is around the 100 g/day mark and falls progressively up to a stool weight of about 200 g/day. Above this, cancer risk is fairly low and does not seem to be affected by changes in stool weight. Whilst the amount of stool passed is of itself clearly not the cause of bowel cancer, it must reflect events within the large intestine which favour the development of neoplasia.

Overall, therefore, there is a relationship between low stool weight and risk of bowel disease. Figures 7.3 and 7.4 each show, in addition to the data on disease, the range of bowel habits for the UK. On average in the UK stool weight is 106 g/day with 47% of the population lying below 100 g/day. It is clear from Figs 7.3–7.5 that the risk of constipation, diverticular disease and bowel cancer increases substantially as bowel habit falls below 100 g/day. There may well, therefore, be a case for trying to increase the population average above this potential danger zone.

7.4 DIET AND BOWEL HABIT

What are the major controls of bowel habit and can anything be done to alter it? The principal factors controlling stool weight in man are diet and transit time. Transit time seems to be an inherited characteristic and little is known of the factors which control it. Although manipulations of diet can be shown to shorten or lengthen transit, there are many factors which we do not know in relation to its control (Cummings, 1978). As far as diet and bowel habit are concerned, NSP stands out from all other components as being fundamental to the control of the amount of stool passed. Many studies have now been performed which confirm this relationship (Cummings, 1986) and no other dietary component has anything like such a major effect. The one exception to this might be RS, although at the present time there are virtually no data to quantitate this. There is a linear relationship over a range of 4–32 g/day NSP, which covers the intake in most countries of the world.

If NSP is a major determinant of stool weight and stool weight is in turn a risk factor of bowel disease, it should be possible to show that NSP intakes are also related to bowel disease risk. Unfortunately, few studies have been done looking specifically at this hypothesis, largely for the reasons given at the beginning of this chapter. However, in the IARC Scandinavian studies, NSP intakes were inversely related to bowel cancer risk (IARC Large Bowel Cancer Group, 1982) and dietary fibre has been shown to protect against bowel cancer in other population and analytic studies (Tuyns *et al.*, 1987; Rosen *et al.*, 1988). Diverticular disease is also less common in populations on diets characterized by high NSP intakes (Gear *et al.*, 1979; Manousos *et al.*, 1985) and clearly constipation can be prevented and on occasions treated successfully by diets containing high NSP. In all these studies, however, it is wisest to say that NSP intake on its own is not the only factor to relate to bowel disease risk, but rather diets which are characterized by high or low NSP intake.

7.5 A RECOMMENDED INTAKE OF DIETARY NSP

Non-starch polysaccharide intake in the UK is about 12.5 g/day (MAFF, 1990). This is found with a stool weight of around 100–110 g/day, which is the population average suggested by other studies. At this stool weight bowel cancer risk is relatively high in the UK and ideally one would wish to shift the population stool weight to something above 150 g/day. This would require an increase in NSP intake from 12.5 to 21 g/day. This is a very substantial change in the UK diet and probably unachievable in the foreseeable future. It may not be wholly necessary either since high NSP diets are also characterized by being high in other complex carbohydrates, particularly RS. They will also be low in saturated fat, which is itself a risk factor for Western diseases.

An increase therefore in NSP intake of around 50% above present levels could be justified on the bowel habit data alone. This would mean raising UK intakes to about 18 g/day, which is of the same order of magnitude as was recommended by the National Advisory Committee on Nutrition Education (NACNE) almost ten years ago, although the actual numbers are different because different methods for measuring dietary fibre were used at that time (Health Education Council, 1983).

Overall, therefore, we would propose that for the UK diets characterized by foods which contain on average 18 g/day NSP should lead to significant changes in bowel habit and thus reduce the risk of bowel disease, particularly constipation, diverticular disease and bowel cancer. Such a change in diet may also be accompanied by other health benefits not directly related to the large bowel.

REFERENCES

Anderson, J.W., Deakins, D.A. and Bridges, S.R. (1990) Soluble fiber; hypocholesterolemic effects and proposed mechanisms, in *Dietary Fiber – Chemistry, Physiology and Health Effects* (eds D. Kritchevsky, C. Bonfield and J.W. Anderson), Plenum, New York, pp. 339–63.

Bass, P. and Dennis, S. (1981) The laxative effects of lactulose in normal and constipated subjects. *J. Clin. Gastroenterol.*, **3**, 23–8.

British Nutrition Foundation Task Force (1990) *Complex Carbohydrates in Foods*, Chapman and Hall, London.

Brodribb, A.J.M. and Humphreys, D.M. (1976) Diverticular disease: three studies. I. Relation to other disorders and fibre intake. II. Treatment with bran. III. Metabolic effects of bran in patients with diverticular disease. *Br. Med. J.*, **1**, 424.

Burkitt, D.P. (1969) Related disease – related cause. *Lancet*, **ii**, 1229–31.

Burkitt, D.P. (1971) Epidemiology of cancer of the colon and rectum. *Cancer*, **28**, 3–13.

Connell, A.M., Hinton, C., Irvine, G., *et al.* (1965) Variation of bowel habit in two population samples. *Br. Med. J.*, **2**, 1095–9.

Cowgill G.R. and Anderson, W.E. (1932) Laxative effects of wheat bran and 'washed bran' in healthy men. A comparative study. *JAMA*, **98**, 1866–75.

Cowgill, G.R. and Sullivan, A.J. (1933) Further studies on the use of wheat bran as a laxative. Observations on patients. *JAMA*, **100**, 795–802.

Cummings, J.H. (1978) Diet and transit through the gut. *J. Plant Foods*, **3**, 83–95.

Cummings, J.H. (1986) The effect of dietary fiber on fecal weight and composition, *Handbook of Dietary Fiber in Human Nutrition* (ed. G.A. Spiller), CRC Press, Florida, pp. 211–80.

Cummings, J.H., Branch, W.J., Bjerrum, L., *et al.* (1982) Colon cancer and large bowel function in Denmark and Finland. *Nutr. Cancer*, **4**, 61–6.

Eastwood, M.A., Smith, A.N., Brydon, W.G. and Pritchard, J. (1978a) Colonic function in patients with diverticular disease. *Lancet*, **i**, 1181–2.

Eastwood, M.A., Smith, A.N., Brydon, W.G. and Pritchard, J. (1978b) Comparison of bran, ispaghula and lactulose on colon function in diverticular disease. *Gut*, **19**, 1144–7.

Findlay, J.M., Smith, A.N., Mitchell, W.D., *et al.* (1974) Effects of unprocessed bran on colon function in normal subjects and in diverticular disease. *Lancet*, **i**, 146–9.

Gear, J.S.S., Ware, A., Fursdon, P., *et al.* (1979) Symptomless diverticular disease and intake of dietary fibre. *Lancet*, **i**, 511–14.

Goy, J.A.E., Eastwood, M.A., Mitchell, W.D., *et al.* (1976) Fecal characteristics in the irritable bowel syndrome and diverticular disease. *Am. J. Clin. Nutr.*, **29**, 1480–4.

Graham, D.Y., Moser, S.E. and Estes, M.K. (1982) The effect of bran on bowel function in constipation. *Am. J. Gastroenterol.*, **77**, 599–603.

Health Education Council (1983) Proposals for nutritional guidelines for health education in Britain. *NACNE*, September.

Hill, M.J., Crowther, J.S., Drasar, B.S., *et al.* (1971) Bacteria and aetiology of cancer of the large bowel. *Lancet*, **i**, 95–100.

Hill, M.J., Drasar, B.S., Williams, R.E.O., *et al.* (1975) Faecal bile-acids and clostridia in patients with cancer of the large bowel. *Lancet*, **i**, 535–9.

IARC Large Bowel Cancer Group (1982) Second IARC International Collaborative Study on diet and large bowel cancer in Denmark and Finland. *Nutr. Cancer*, **4**, 3–79.

Kromhout, D., Bosschieter, E.B. and Coulander, C.L. (1982) Dietary fibre and 10-year mortality from coronary heart disease, cancer and all causes. The Zutphen study. *Lancet*, **ii**, 518–22.

Kushi, L.H., Lew, R.A., Stare, F.J., *et al.* (1985) Diet and 20-year mortality from coronary heart disease. The Ireland-Boston Diet-Heart study. *N. Engl. J. Med.*, **312**, 811–18.

MAFF (1990) *Annual Report of the National Food Survey*, HM Stationery Office, London.

Manousos, O., Day, N.E., Tzonou, A., *et al.* (1985) Diet and other factors in the aetiology of diverticulosis: an epidemiological study in Greece. *Gut*, **26**, 544–9.

Marcus, S.N. and Heaton, K.W. (1986) Intestinal transit, deoxycholic acid and the cholesterol saturation of bile – three inter-related factors. *Gut*, **27**, 550–8.

Marcus, S.N. and Heaton, K.W. (1987) Irritable bowel-type symptoms in spontaneous and induced constipation. *Gut*, **28**, 156–9.

Marlett, J.A., Li, B.U.K., Patrow, C.J. and Bass, P. (1987) Comparative laxation of psyllium with and without senna in an ambulatory constipated population. *Am. J. Gastroenterol.*, **82**, 333–7.

Meyer, F. and Le Quintrec, Y. (1981) Rapports entre fibres alimentaires et constipation. *Nouv. Presse Med.*, **10**, 2479–81.

Moore-Gillon, V. (1984) Constipation: what does the patient mean? *J. R. Soc. Med.*, **77**, 108–10.

Morris, J.N., Marr, J.W. and Clayton, D.G. (1977) Diet and heart: a postscript. *Br. Med. J.*, **2**, 1307–14.

Ornstein, M.H., Littlewood, E.R., McLean Baird, I., *et al.* (1981) Are fibre supplements really necessary in diverticular disease of the colon? A controlled clinical trial. *Br. Med. J.*, **282**, 1353–6.

Painter, N.S. (1975) *Diverticular Disease of the Colon – a Deficiency Disease of Western Civilization*, William Heinemann, London.

Parks, T.G. (1974) The effects of low and high residue diets on the rate and transit and composition of the faeces. *Proc. 4th International Symposium on Gastrointestinal Motility, Banff, Alberta, Canada, 1973*. Mitchell Press, Vancouver, pp. 369–80.

Rosen, M., Nystrom. L. and Wall, S. (1988) Diet and cancer mortality in the counties of Sweden. *Am. J. Epidemiol.*, **127**, 42–9.

Royal College of General Practitioners OPCS/DHSS (1986) Morbidity statistics from general practice – third national study, 1981–1982, Series MB5 No. 1, HM Stationery Office, London.

Royal College of Physicians (1980) *Medical Aspects of Dietary Fibre*, Pitman Medical, Tunbridge Wells.

Sandler, R.S., Jordan, M.C. and Shelton, B.J. (1990) Demographic and dietary determinants of constipation. *Am. J. Public Health*, **80**, 185–9.

Segal, I., Solomon, A. and Hunt, J.A. (1977) Emergence of diverticular disease in the urban South African black. *Gastroenterology*, **72**, 215.

Setchell, K.D.R., Street, J.M. and Sjovall, J. (1987) Faecal bile acids, in *The Bile Acids* (eds K.D.R. Setchell, D. Kritchevsky, and P.P. Nair), Plenum, New York, pp. 441–571.

Smith, A.N., Drummond, E. and Eastwood, M.A. (1981) The effect of coarse and fine Canadian Red Spring Wheat and French Soft Wheat bran on colonic motility in patients with diverticular disease. *Am. J. Clin. Nutr.*, **34**, 2460–3.

Smith, R.G., Rowe, M.J., Smith, A.N., *et al.* (1980) A study of bulking agents in elderly patients. *Age Ageing*, **9**, 267–71.

Sonnenberg, A. and Koch, T.R. (1989) Physician visits in the United States for constipation: 1958–1986. *Dig. Dis. Sci.*, **34**, 606–11.

Tarpila, S., Miettinen, T.A. and Metsaranta, L. (1978) Effects of bran on serum cholesterol, faecal mass, fat, bile acids and neutral sterols, and biliary lipids in patients with diverticular disease. *Gut*, **19**, 137.

Taylor, I. and Duthie, H.L. (1976) Bran tablets and diverticular disease. *Br. Med. J.*, **1**, 988–90.

Thompson, W.G. and Heaton, K.W. (1980) Functional bowel disorders in apparently healthy people. *Gastroenterology*, **79**, 283–8.

Trowell, H. (1972) Ischemic heart disease and dietary fiber. *Am. J. Clin. Nutr.*, **25**, 926–32.

Tuyns, A.J., Haelterman, M. and Kaaks, R. (1987) Colorectal cancer and the intakes of nutrients: a case control study in Belgium. *Nutr. Cancer*, **10**, 181–6.

Yano, K., Rhoads, G.G., Kagan, A. and Tillotson, J. (1978) Dietary intake and the risk of coronary heart disease in Japanese men living in Hawaii. *Am. J. Clin. Nutr.*, **31**, 1270–9.

8 Dietary fibre: the facts?

I. Macdonald

8.1 INTRODUCTION

It is not an unfamiliar experience to most of us to feel that when someone comes up with a good idea you wish you had thought of it yourself, and partly because of this there is a tendency to accept the fresh notion and to use it to explain many unsolved problems. Such was the response to the concept first postulated by Cleave and Campbell (1966) and later taken up by Burkitt and Trowell (1975), that many of the ills of Western man could be attributed to too much refined carbohydrate in the diet or, on the other side of the coin, too little 'roughage' or fibre. It was a brilliant idea that was immediately accepted and also an unorthodox idea, in that a low intake of a non-nutrient could shorten our lives.

8.2 ASSOCIATION OR CAUSE AND EFFECT

Many research studies spawned by this concept were, and are, directed towards the therapeutic role of dietary fibre (we all know what fibre is even though the experts cannot agree about defining or measuring it) and findings from these studies have certainly resulted in the alleviation of suffering. About this aspect of dietary fibre there can be no quarrel. It is with the so-called preventive role of dietary fibre that there is some doubt and it is, of course, easy to have this doubt because it is a well nigh impossible area in which to carry out sound scientific research. Nevertheless, this fact should not be the excuse for accepting hypotheses as if they were proven. Hypotheses in nutrition and other biological areas may arise by the association, positive or negative, of two facts. A somewhat ridiculous example is that as the sales of nylon stockings have increased so has the incidence of coronary heart disease. This is a positive association with no doubt a high p value, but it is not a causal relationship. Essentially, association of facts is what much of epidemiology is about and this is the main basis of the fibre hypothesis, when applied to the aetiology of disease. There is no doubting the negative association between the amount of fibre consumed and the incidence of diseases such as obesity, coronary heart disease, diabetes, diverticulosis, etc., but does this mean a cause and effect relationship?

The next and much more difficult step is to determine whether, in fact, this association is causal, that is whether X is related to Y or just associated with it. Incidentally, if epidemiology does not show an association between two variables then a causal relationship is highly unlikely. Another pitfall in reasoning, that is common in nutrition, is to equate the successful treatment of a condition by a known substance with the cause of the condition being due to a deficiency of that substance. It is almost a conditioned reflex for the nutritionist because he or she has considerable experience with this type of metabolic reasoning, examples of which are vitamins in vitamin-deficient states, iron and anaemia, protein and malnutrition, etc. Although dietary fibre may alleviate the symptoms of diverticulosis its deficiency in the diet may not be the cause of the condition. As Dr Eastwood once remarked, 'aspirin may cure a headache but a headache is not due to aspirin deficiency'. Before considering some pathological conditions *vis à vis* the fibre hypothesis in more detail, there are some more general points that give reason to doubt the fibre hypothesis.

8.3 DIETARY INTAKE

It is well known how difficult it is to assess with any degree of accuracy the dietary intake of a person, or indeed a group of people, especially in some parts of the world where it is claimed that dietary fibre intake is high (e.g. Africa). It seems that in the more developed countries not only is total dietary intake an unreliable figure, but the inhabitants may well have had a higher fibre intake than previously considered, if dietary fibre was defined as 'the sum of lignin and the plant polysaccharides which are not digested by the endogenous secretions of the human gastro-intestinal tract' (Trowell, 1974). I refer, of course, to 'resistant' starch. It has been remarked that in the UK if resistant starch is included, our dietary fibre intake may not be much different from that of the Bantu in South Africa (Cummings, personal communication). If this is true then the general hypothesis that dietary fibre has a preventive role becomes very uncertain.

Not only may values for dietary intake in general be suspect, so also may the figures for morbidity and mortality be unreliable. The low incidence of colonic diverticula in the East (where seemingly there is a low fibre intake) has been challenged and futhermore the precise cause of sickness and death may well be more accurate in the West with its more modern diagnostic methods. Also, as has been mentioned by others, death occurs at an earlier age in the developing world before, perhaps, the age at which colon cancer, etc., commonly develops.

It is true that few epidemiological studies have refuted the dietary fibre hypothesis, but these studies have not assessed the amount or type of fibre in the diet, only the overall diet and it may well be that it is not the fibre in the diet that is beneficial, but some other constituent of the diet that accompanies the dietary fibre. There is some evidence from experimental animals that vegetables, which of course tend to have higher fibre levels, contain anti-cancer agents (Wattenberg, 1983) as well as fibre. It has been reported that in the USA there is an inverse relationship, epidemiologically, between the incidence of colon cancer and the amount of sunshine. Peoples with high fibre diets in Africa tend to live in sunny climes. These are not alternative hypotheses, but examples of the many uncontrolled variables in the evidence that postulates that a high fibre intake reduces the prevalence of certain diseases common in the Western world.

8.4 DIETARY FIBRE AND THE COLON

It has been implied that the greater the dietary fibre intake the shorter the gut transit time, although this has also been challenged. In some people the transit time may not be affected by dietary fibre (Findlay *et al.*, 1974) although the stool consistency usually is. Futhermore, there is an assumption made here that there is a physiological transit time and that we should be within that range of time. It seems that there is no evidence to support this view, and the concept that defaecation must be a daily event is surely without scientific support. It is true that when it occurs defaecation should not be subject to excessive straining, and constipation is to be avoided. It is also true that the call to defaecate should be obeyed, but is there any evidence to suppose that it is harmful in the short or long term if the desire to defaecate occurs only every 2, 3 or 4 days?

Dietary fibre vies with dietary fat as the food item whose deficiency or excess, respectively, is supposed to cause colon cancer. The current view from the experts seems to be that as far as fibre is concerned the case is not proven. Presumably low fibre in the colon is not, *per se*, a cause of cancer, but its absence may allow carcinogens to act without the protection of fibre. Also a high fibre diet may mean a reduced intake of fat, which may be carcinogenic, and a high fibre diet, as mentioned earlier, may allow a greater intake of anti-carcinogens (Wattenberg, 1983). The US Bureau of Foods pointed out in 1980 that the association of low fibre intake with diverticular disease, and colon cancer, is faulted by genetic, environmental, cultural, dietary and other uncontrolled variables and this statement still applies.

8.5 GALL STONES

Cleave and Campbell (1966) in their book on the saccharine diseases which led to the fibre hypothesis, stated that too much refined carbohydrate and therefore not enough fibre in the diet was responsible for the formation of gall stones. However, the Professor of Medicine at Edinburgh University (Bouchier, 1990) stated that 'No evidence exists that any particular diet influences gall or gall bladder disease' and 'that a high fibre diet does not prevent the recurrence of gall stones'.

8.6 OBESITY

In his early publications on the dietary fibre hypothesis, Trowell (1976) argued that in Africa, where he spent most of his clinical life, he saw no obesity except in urban populations where more Westernized (and, he presumed, low fibre) food was consumed. It requires little imagination to realize that the rural population in Africa has a reduced food intake, whether fibre enriched or not, and therefore for this reason the people are very unlikely to be obese. Moreover, it is entirely plausible from the comfort of an armchair to postulate that dietary fibre will prevent obesity for these seemingly obvious reasons:

1. fibre reduces the energy density of a food;
2. fibre reduces the rate of absorption of food energy;
3. fibre, by virtue of its extra volume, induces satiety.

The potential role of fibre-rich foods or supplements in preventing or treating obesity, despite the theoretical reasons why it should be useful, is still open to question. The report of a committee set up by the Canadian Health and Welfare Department (1985) stated

> although in the final analysis obesity is related to a disturbance in energy balance, the mechanisms by which this process occurs or could be reversed by dietary fibres are not clear. In short term studies, fibre-rich diets or supplements (fruits, vegetables, whole grains and probably guargum) have been shown to achieve an increased satiety (reported subjectively) and in a very few studies, to contribute to modest weight loss. There is little information on whether these high fibre diets or fibre preparations would be acceptable for long term use. The evidence to date is in no way sufficient either to establish weight reduction as a physiological effect of fibres, or to determine the role of fibres in weight loss preparations.

In 1978 a workshop on 'Fiber and Obesity', sponsored by the US National Institutes of Health, made the following recommendations.

1. The hypothesis that a fibre-rich diet may be an effective obstacle to obesity needs testing by means of controlled experiments in animals and man.
2. These experiments should take into account such factors as (a) the low fat content of fibre-rich foods, (b) the effects of such manipulations on palatability, ingestion rate, effort of chewing, intestinal bulk, etc.
3. The studies should be of sufficient duration to distinguish between transient and sustained effects on food intake and body weight.

4. Various types of food fibre should be tested for their ability to act as an obstacle to excessive energy intake.

This point of dietary fibre and body weight loss has been emphasized because it shows very well how an apparently watertight hypothesis, with seemingly good theoretical reasoning to back it up, may indeed be invalid because there are so many uncontrolled variables, each of which, either alone or in concert with another variable, may be responsible for any weight loss that occurs, and even this fact has yet to be established.

8.7 CONCLUSION

The intention in this treatise is not to dismiss the dietary fibre hypothesis in the causation of disease, far from it, but it is important to point out that the evidence does not yet exist, that would raise an association into a causal relationship in this hypothesis. Finally, it is also important to bear in mind that dietary fibre does have some accepted disadvantages and that the young, the old and the pregnant should perhaps be circumspect in their dietary fibre intake.

REFERENCES

Bouchier, I.A.D. (1990) Gall stones. *Br. Med. J.*, **300**, 592–6.

Burkitt, D.P. and Trowell, H.C. (1975) *Refined Carbohydrate Foods and Disease. Some Implications of Dietary Fibre*. Academic Press, London.

Canadian Health and Welfare Department (1985) *Report on Dietary Fibre*, Government of Canada, Ottowa, p. 20.

Cleave, T.L. and Campbell, G.D. (1966) *Diabetes, Coronary Thrombosis and the Saccharine Disease*. Wright, Bristol.

Findlay, J.M., Michell, W.D., Smith, A.N., *et al.* (1974) Effects of unprocessed bran on colon function in normal subjects and in diverticular disease. *Lancet*, **i**, 146–9.

National Institutes of Health (USA) Workshop (1978) Fiber and obesity – summary and recommendations. *Am. J. Clin. Nutr.*, **31**, S252–4.

Trowell, H.C. (1974) Definitions of fibre. *Lancet*, **i**, 503–5.

Trowell, H.C. (1976) Definitions of dietary fiber and hypotheses that it is a protective factor in certain diseases. *Am. J. Clin. Nutr.*, **29**, 417–27.

US Bureau of Foods (1980) *The Role of Dietary Fiber in Diverticular Disease and Colon Cancer*, Federation of American Societies of Experimental Biology, USA.

Wattenberg, L.W. (1983) Inhibition of neoplasia by minor dietary constituents. *Cancer Res.*, **43**, 2448–53.

DISCUSSION

Rosenberg, Massachusetts I am fascinated by Dr Cumming's approach to quantify fibre intake, but I am concerned with the data base and I expect you are because of the difficulty you had with collecting it. First of all as a gastroenterologist, I have some problems with the simple definition of the quantification of bowel habit as stool weight. It would not be the marker that I would think to be the most likely to be related to the pathogenesis of colon cancer. But even if that were a good technique, if I understand the data correctly, the measurements that were made of stool weight in diverticular disease and in constipation were no different than the measurements made in the non-involved population. These were not markers of diverticular disease in terms of stool weight, they were simply markers of the population that were being measured. If that is the case then how can one use those numbers as guides for achievement of a certain dietary intake in order to prevent disease?

Cummings I chose stool weight as a risk factor because it is the one where there is most information available, but I quite agree that if I had to set up a risk factor profile for bowel cancer I would have certain other things in it. I would want to know about faecal butyrate amongst other things. I picked stool weight as something that everybody would know about and for which there was a moderate amount of information, whereas I could not find, looking in the literature, any other useful marker for bowel disease that I could correlate with these conditions. So it is the best in the present circumstances. Yes, it is surprising that the average stool weight of people who are constipated and have diverticular disease are so close to the population average of the UK, but one has to face the fact that bowel symptoms are very common and affect a major proportion of the UK population, particularly the elderly. There is a difference of about 40 g per day between the average for constipated people and that of the whole population.

Rosenberg Is there actually a significant difference between the population with diverticular disease and the matched population without diverticular disease in stool weight?

Cummings Nobody has ever done that sort of study, of course, because very few people ever collect stools for understandable reasons. The diverticular disease data include people with diarrhoea.

Kritchevsky There is the study done in Hawaii where they compared bowel habits and colon cancer in Japanese who had moved to Hawaii when they were adolescents or adults, in their children who were born in Hawaii and in Caucasians. The Japanese fathers and sons had the same transit times and stool weight. Transit time was much lower than that of the Caucasians, the stool weight much higher. The Japanese fathers had a very low colon cancer rate and the sons had the same incidence as the Caucasians and so that is one example where the particular combination did not give a difference.

Eastwood, Edinburgh It would have been interesting if you had been able to show our data for Scotland which has a much lower dietary fibre intake than in Cambridge or Bristol. It is possible that distinct differences would have been brought out. Diverticulosis is a very complex condition to look at because to my mind the development of diverticulosis is part of the ageing process which may or may not be magnified by the amount of dietary fibre. Separate from this is symptomatic diverticulosis which develops when people have a low stool weight, experience abdominal pain and pass pellety stools. There are many individuals with diverticulosis who are asymptomatic and as far as stool weight is concerned are indistinguishable from the normal population. So that in terms of stool weight there is a distinction between symptomatic patients and asymptomatic individuals with diverticulosis.

Cummings There are good precedents for the ranges of a particular variable overlapping between the normal population and the disease population. A classic example is acid secretion in peptic ulcer where there is a huge overlap between the duodenal ulcer patient and the healthy controls, but nobody would deny that acid has a major role to play in peptic ulcer because you can treat it so effectively by switching off acid secretion with H_2 receptor blockers, etc. So, this business with overlap is a problem and clearly there are other factors at work and there may be better risk markers if we can develop them. I put this up as a first post, or a first marker for discussion.

Hewitt, Nottingham Can I widen the discussion for a moment and ask for your comment about the Masai warrior who lives on a fibre-free diet of blood and milk and looks disgustingly healthy on it; and possibly bring in the meal replacement diet, that say the astronaut has, which is a zero-residue diet, presumably for reasons of practicality. Is that actually a risk to health having no fibre there at all?

Cummings Well as you know the astronauts gave up these fibre-free diets because they gave them so much bellyache. On a fibre-free diet stool weight falls to about 40 g a day. Without fibre in your diet, stool weights are really very low and the astronauts' experience suggested that this was not a practical way to spend long times in space – straining at stool. The Masai: it is a total misnomer that the Masai do not eat vegetables, it is just that they are very seasonal. Is that right? It is only certain times of the year when they have the blood and certainly I do not think the women drink the blood at all. They live entirely on the vegetables.

Hewitt I believe it is the warriors only who get the blood.

Plummer, Cambridge I would like to ask about the very clear differences between women and men in constipation? Do women have more bowel disease? Some women suggest that men drink a lot of beer and that is why they don't get constipated. This is local folklore. What are your comments?

Cummings It is a frequently observed finding that women have slower transit times and lower stool weights than men in most studies, but not all. It is pretty universal in western societies at any rate that women have smaller stool weights and slower transit. Transit is the key thing in determining stool weight. So why do women have slower transit time and what controls it? The glib answer is that it is progesterone which is affecting smooth muscle in the larger intestine. I am not sure it is an entire explanation, but this is an area where very little research has been done.

Fell, Glasgow Would Dr Cummings like to speculate on the possible harmful effect of increasing fibre on the bioavailability of other dietary components which are essential, particularly calcium and zinc which may be, some people consider, in marginal supply in the general population anyway. So if you add extra wheat bran or some other source will you not push large numbers of the population over into frank deficiency states.

Cummings The answer is no. The increase I am suggesting, going from 13 to 18 g a day, is modest in terms of global intakes. I would not wish to see the British population doing this solely by eating wheatbran. I think it has to come from a change in diet characterized by an NSP intake of 18 g a day.

Kritchevsky I think studies in the United States show that for a well-fed population, even when the amount of fibre in the American diet was almost doubled, it did not affect trace mineral excretion. There was one experiment in which they did show increased calcium excretion and then remembered that they put spinach into the diet.

Cliff, Food and Health Research A question for Dr Cummings. At one stage in your talk you asked the rhetorical question; how do we achieve these targets which you have set? You did not say a great deal about that, but in the COMA report on bread and flour, which you were involved in writing almost a decade ago, you had the most extensive catalogue of practical recommendations for increasing fibre consumption, more practical recommendations than any other COMA report written this decade. Now 10 years later, how would you suggest on reflection we shoud help people increase their fibre consumption as a practical matter?

Cummings I think this is a question for the British Dietetic Association. How you change the diet of the population is a difficult problem and one that needs to be addressed by experts who know what the population is eating, what determines its habits and how you can change it. We all know where the NSP in the diet is, how you change it is a major problem.

Mela, Reading There was a strangely reasonable sounding hypothesis put forward, about 10, 15 years ago, that diverticular disease in developed nations results primarily from gas pressure in the colon, stemming from culturally reinforced flatus retention. Since we know that high-fibre foods are generally recognized as being windy, could it be that without a concomitant change in social acceptance of flatulence, in fact a high-fibre diet could be counter productive *vis à vis* diverticular disease.

Lund, Norwich When you feed resistant starch you have much less affect on faecal bulking, is that because the resistant starch is being fermented, and the bulking effect is purely that of increased bacteria?

Cummings It is too early in our studies for me to answer that question.

Reid, London John Cummings has said going from 13 to 18 g is a small increase, it is about 40%. Do you have a comment on that?

Cummings I cannot argue with the calculation. It could be said that it is a conservative increase because you could possibly make a case for it being larger if you put the UK bowel habit in context with the rest of the world. If I put up a national league table, we, and particularly Scotland, would be right down at the lower part of the table.

Barker, Southampton Could I just question the history of the fibre hypothesis? My understanding is that the fibre hypothesis arose from the paper by Professor Rendal

Short in 1920 who concluded that the origins of appendicitis lay in the reduced amount of cellulose being eaten. There was a leading article in the *Lancet* about dietary fibre in 1927, and Dennis Burkitt always acknowledged that Professor Rendal Short was the originator of the hypothesis. Now the problem about appendicitis being the origins of the hypothesis is that it does not withstand critical examination, for example the fibre hypothesis cannot explain the progressive decline in appendicitis rates during the past 40 years, in most Western countries. The data against fibre are so persuasive that even Dennis Burkitt now admits that dietary fibre, whatever else it does, does not explain appendicitis and there are better ideas around. Now having undermined the origins of the dietary fibre hypothesis, this does not of course bear on other possible applications of it and I have no quarrel at all with what John Cummings said. My quarrel really is that Professor Macdonald seems to be laying the blame at the door of epidemiologists. I suggest to him that it was not the epidemiologists who greeted this with great enthusiasm, it was people who felt that the mechanisms were rather enticing and worthy of exploration.

Macdonald Epidemiology provided what they thought was evidence for their hypothesis. I do not think their evidence as shown on my slide was very good.

Kritchevsky Don't you think some of the public enthusiasm is because for the first time the public were told by the medical profession they could add something to the diet instead of taking something out? You know, I think there's a lot less resistance to that sort of thing.

Bender, London On the question of how to get people to increase the amount of dietary fibre, there is quite a lot of evidence that it is difficult. There was a closely controlled study of 48 people in which they were given advice by the dietician as to what to do with their diet and they all succeeded in lowering the fat, increasing the PS ratio and so forth, but not in increasing the dietary fibre. With reference to the work by Louise Davis on the elderly in which she rather despairingly stated that she could not get the elderly people to increase their dietary fibre. When I said to her 'Well did it matter, did they need to, were they suffering from constipation?' she answered 'Oh, no there was nothing wrong with them, but of course in improving their diet we had to improve the dietary fibre'. Just one last point, on this question of beer, one of the richest sources of dietary fibre in the world is Guinness, which is 0.5% pentosans, and you do not drink Guinness by the 100 ml, you drink it by the litre.

Cummings Beer also has a lot of sulphate, which is laxative in reasonable quantities, which may partly explain the action of beer.

Claret, Glasgow Would Dr Cummings care to give an opinion on the amounts of fibre for children?

Cummings We must be very cautions about saying anything about infants having fibre, and for children I think they should have similar amounts to adults, but reduced in relation to energy intake.

Kritchevsky There was an American group that were discussing fibre intake recommendations. One recommendation was that fibre intake should be geared to caloric intake or to energy intake – grammes per thousand kilocalories. As far as trace minerals are concerned, it is the very young and the very old that may suffer from excessive fibre intake.

Cummings There is very, very little information on the effect of fibre in young people.

Speedy, Glasgow A food and health policy was introduced in the Neasden Hospital and the Brent Area in about 1985/86. Elderly long-stay patients were given dietary fibre advice based on the NACNE recommendations. The older patients enjoyed the dietary experiment, and the hospital had saved so much money from laxatives and aperients that they were able to transfer finances from the pharmacy budget to the catering budget. I interviewed some of the older patients myself who said that they had felt an awful lot better since they had had all the lovely fresh vegetables.

Heaton I think at some stage in this discussion we ought to mention vegetarians. Jill Davies has very detailed and careful studies on some 19 vegetarians and 19 vegans and has shown not only, as you would expect, that their fibre intakes are higher than those of omnivores, but also that their stool weights are also higher and that their transit times are faster. The expected correlations exist in both groups between fibre and bowel function. Vegans are a self-tested group in which, to answer the question that was raised earlier about possible harmful effects of high fibre intakes which are much higher than John Cummings's suggested 18 g a day, there is no evidence that they suffer from deficiency of calcium, iron or zinc or any other metal. The only deficiency they are prone to is vitamin B_{12}, but most of them know about this and can easily avoid it.

Kritchevsky In the United States there is the population of Seventh Day Adventists, a group that will soon have been studied more than the Wistar rat. About 3% of them are true vegetarians or vegans, about 50% are lacto-ovo-vegetarians, the rest eat meat sparingly. We have looked at the cholesterol concentrations in those three groups, as well as in the general population, and it turned out that really low cholesterol was only found in the vegans. Once you breach the absolute vegetarian style, the cholesterol level was almost as high as that of the general population. In general, an analysis of their diets showed the only real difference was the pectin content of their diets which was high in the vegan diet.

Again there are problems with making comparisons, because at one time this group's low incidence of colon cancer and breast cancer compared to the rest of the American population was ascribed to their modest meat intake. Then somebody violated the first law of research, which is that you never repeat a successful experiment, and looked at Mormons, who eat as much meat as anybody in the United States. They had the same incidence of breast cancer as the Seventh Day Adventists and even less colon cancer, so there is still something here that requires a little bit of interpretation.

PART FOUR

Salt

9 Salt and hypertension: the controversy continues

D.G. Beevers

9.1 INTRODUCTION

The hypothesis that dietary salt intake is related to the blood pressure remains controversial. It is based on clinical, epidemiological and laboratory evidence. Whilst many studies provide strong support for the salt hypothesis, there are also a great many negative or unconvincing reports, and it is likely that the final proof will never become available. It behoves all experts in the field, therefore, to examine the data dispassionately and avoid the temptation to overstate their case, either in favour or against the concept.

9.2 ANCIENT TIMES

The first recorded observation that salt and blood pressure are related appears in a textbook of medicine written 4000 years ago in ancient China. Huang Ti, the Yellow Emperor, stated 'if too much salt is used in food, the pulse hardens'. This may well be a description of a full volume pulse of a patient with hypertension; the Chinese were, at that time, very expert in the interpretation of the pulses.

9.3 EARLY CLINICAL DATA

In more recent times, there have been clinical studies which strongly suggest that salt and blood pressure are related. Ambard and Beaujard (1904) recommended a low salt diet for people with high blood pressure and subsequently Kempner (1944) introduced a rice/fruit diet which contained no sodium. Whilst these diets were intolerable in the long term, their use could be justified for very severe hypertensives, at a time when no effective antihypertensive drugs were available. When the thiazide diuretics and other powerful drugs were introduced, the use of no-salt diets to control blood pressure fell into disrepute.

9.4 LABORATORY EVIDENCE

Laboratory evidence in favour of the salt hypothesis comes from studies of specifically bred strains of rats who are either salt sensitive or salt resistant (Knudsen and Dahl, 1966). Salt sensitive rats develop a marked rise in blood pressure with advancing age and have a high incidence of stroke. By contrast, salt resistant rats do not develop high blood pressures. It has been pointed out, however, that the amount of salt that is needed to induce hypertension in rats is, on a weight for weight basis, a great deal higher than that which is ever seen in humans.

9.5 EPIDEMIOLOGY

Epidemiological evidence in favour of the salt hypothesis comes largely from studies comparing primitive rural populations with advanced westernized societies. It was noted that primitive populations mainly in rural Africa or South America showed no rise in blood pressure with advancing age and, as a consequence, hypertension was practically non-existant. Many of these societies consumed very small quantities of salt in their diet. Subsequently Louis Dahl (1961) was able to collate data from a large number of population surveys and show that populations who eat more salt have higher blood pressures and he reported a positive correlation between salt intake and blood pressure.

There were many objections to Dahl's work. Most of the population studies that he included in his overview analysis were conducted at different times, by different observers and whilst using differing criteria for the diagnosis of hypertension. Blood pressures were measured with unstandardized methods and major confounding variables like body mass index and alcohol intake as well as the age of examinees was not taken into account.

9.6 THE INTERSALT PROJECT

A major advance in research into the salt hypothesis was provided by the recent publication of the INTERSALT project (INTERSALT Cooperative Research Group, 1988). This study had its origins in a 10-day residential seminar in advanced cardiovascular epidemiology sponsored by the International Society and Federation of Cardiology which took place in Tuohilampi, Finland in 1982 (Beevers, 1989). The fellows attending the seminar (each representing a different country) were given the task of devising a project which would answer the very serious epidemiological objections to Louis Dahl's international comparisons. Theoretical plans were formulated on the best methods of conducting a truly international collaborative project, which would be feasible in both urban populations and remote rural communities. On the last day of the seminar the fellows presented their ideas and at this stage it was suggested by the course organizers that this should no longer be a paper exercise but that the project should actually be conducted. Thus was born the INTERSALT project.

In its final form, the INTERSALT project was a study of blood pressure in relation to 24-hour urinary sodium excretion (as a marker for salt intake) in 52 populations in 32 countries. This included communities from India, continental China, the Soviet Union, the USA, Africa and South America as well as Europe and the UK. Strenuous efforts were made to measure all possible factors that might be related to the development of high blood pressure. There is a known association between blood pressure and obesity, high alcohol intake, advancing age and concurrent diseases and ambient temperature and all these factors had to be taken into account. All clinical and laboratory methods were rigorously standardized and all observers, no matter how senior or experienced, were required to attend special training sessions. All blood pressures were measured with a Hawksley random zero sphygmomanometer and all urine samples were analysed at a single central laboratory in Leuven, Belgium.

The INTERSALT project was published finally in 1988 and it confirmed unequivocally that there was indeed a relationship between urinary sodium excretion and blood pressure, although this did not impress some critics (Swales, 1988). The INTERSALT project also showed that the negative association between blood pressure and potassium intake was less significant than had previously been considered likely. Furthermore, the measurement of the sodium/potassium ratio did not provide further information.

It should be stressed, however, that the INTERSALT project also demonstrated that the relationship between blood pressure and both body mass index and alcohol intake was considerably more powerful than the relationship between blood pressure and sodium excretion.

Close analysis of the INTERSALT data demonstrates, however, that the correlation between salt intake and blood pressure depends largely on the four 'primitive' populations from rural Africa and South America. When data from these four centres are removed from the analysis and the remaining 48 centres only are studied, the relationship between salt and blood pressure ceases to be significant. It is doubtful whether the removal of these four centres from the INTERSALT analysis is statistically or methodologically permissible. Research workers should avoid the temptation to include or exclude data from their analyses which do or do not support their preconceived notions. The absence of a relationship between sodium excretion and blood pressure in the analysis of the 48 centres naturally pleases those who have denied the validity of the salt hypothesis. Separate examination of the data from the 52 individual centres, however, shows that 15 were able to demonstrate a significant positive association between salt excretion and blood pressure and only two showed a significant negative association. Thus, at least some of the within-centre correlations (as opposed to the comparison of centres) also lend support to the salt hypothesis.

The INTERSALT project was conducted at a time, however, when at least some national dietary salt intakes were changing rapidly. The northern Japanese populations, which were previously known to have a high salt intake, had already dropped their salt intake with a consequent reduction in stroke mortality rates (Elliott, 1989). The INTERSALT project, unfortunately, did not have enough populations with high salt intakes and only had four populations with very low salt intakes. The remaining populations in Europe and North America were all very much the same.

With or without the four low sodium populations, however, the INTERSALT project does show a highly significant correlation between salt excretion and the relationship between systolic blood pressure and advancing age (Fig. 9.1). These data, therefore, strongly suggest that high salt intake is related to the rise in blood pressure with age, which is steep in westernized societies and absent in primitive societies.

It is unlikely that any further international epidemiological evidence will become available to confirm or deny the salt hypothesis. The INTERSALT project was the biggest ever international epidemiological study and its successful completion reflects the excellent working relationships generated at the International Society and Federation of Cardiology's annual residential teaching seminars on epidemiology. Small-scale projects do, however, provide further evidence in favour.

9.7 MIGRATION STUDIES

One important study has been the Kenya Luo migration project in which rural Kenyans living near lake Victoria have been studied before and then after migrating to live in urban Nairobi (Poulter *et al.*, 1990). Blood pressures of these examinees rose sharply soon after arrival in the city and this occurred in the context of a marked shift in dietary intake of both sodium and potassium. Rural Kenyans eat very little salt but consume a high potassium, vegetable-based diet. By contrast, many Kenyans living in Nairobi eat a high salt, rather more westernized diet. Studies in South Africa have also demonstrated that rural Xhosa tribemen have very low blood pressures, whereas genetically identical people, often in the same families, living in the cities have even higher pressures than the white minority population (Sever *et al.*, 1980). Again, major differences in sodium and potassium intake are likely to be major contributors to this marked urban/rural blood pressure difference.

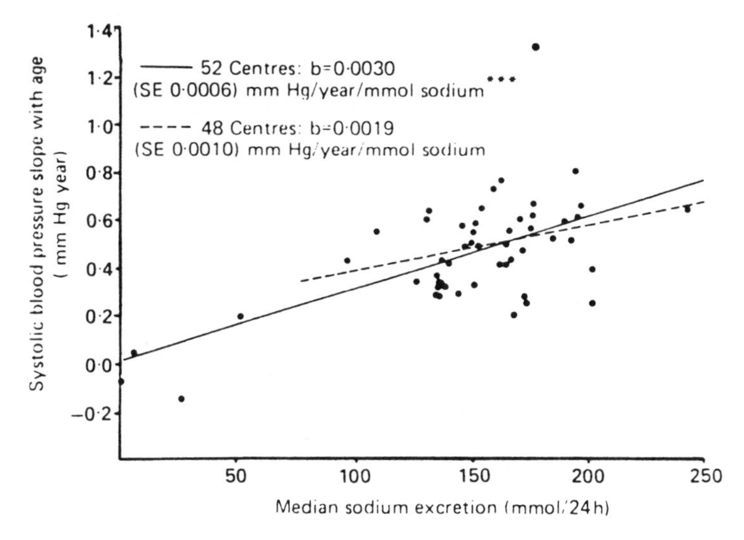

Figure 9.1 Relationship between 24-hour urinary sodium excretion and the slope of systolic blood pressure with advancing age. Data from the INTERSALT study in 52 communities in 32 countries (INTERSALT Cooperative Research Group, 1988). Data are presented both with (continuous line) and without (broken line) the four low salt populations.

9.8 SALT DEPLETION STUDIES

At a clinical level, there is also controversy about the relationship between salt and blood pressure. There have been numerous studies in which hypertensive patients have been subjected to a low salt diet and their blood pressures observed. Many of these studies had methodological problems, often because of the awareness by the patients that they were taking a low salt diet and the consequent effects on their perception of blood pressure. Only a few studies can be regarded as reliable, having overcome some of these problems. Two of these were randomized controlled trials of placebo versus active slow sodium tablets in sodium-deplete, hypertensive patients. Thus, all participants were taking a low salt diet throughout but they received salt supplements or placebo salt supplements in a manner that did not affect their awareness of their salt intake. Of the two studies, one conducted in a blood pressure clinic in London (MacGregor *et al.*, 1982) showed a convincing drop in average blood pressure in response to salt depletion (Fig. 9.2). By contrast, the other study showed no effects on blood pressure. This negative study from a general practice in South Wales was, however, conducted amongst a group of very mild hypertensive individuals who had not previously been considered severe enough to warrant antihypertensive treatment (Watt *et al.*, 1983). There may

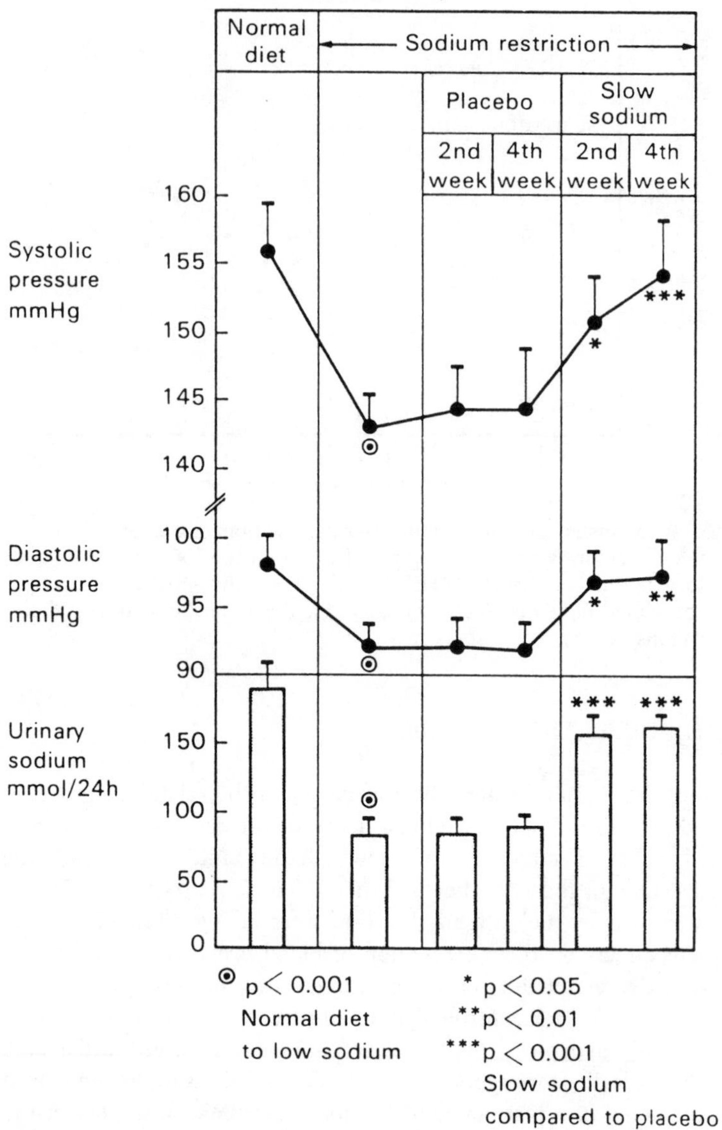

Figure 9.2 Randomized double-blind study of placebo versus active slow sodium tablets in sodium restricted hypertensives (MacGregor *et al.*, 1982).

be no contest between these two studies because it is possible that more severe hypertensive patients may be more salt sensitive than milder cases or normotensive people (MacGregor, 1985). Thus the two studies were studying different populations.

The concept of individual variation of salt sensitivity thus requires further investigation. Just as there are salt sensitive and salt resistant rats, so also there may be salt sensitive and salt resistant people. Close examination of the positive clinical study of sodium depletion in hypertensive patients (MacGregor *et al.*, 1982) does demonstrate that some patients did not respond to their low salt diet. It is possible that not only are hypertensives more salt sensitive than normotensives but also that certain people, including Afro-Caribbean patients in particular, are more salt sensitive than Caucasians.

9.9 SODIUM RESTRICTION IN POPULATIONS

Two community-based, controlled intervention trials of sodium restriction have been conducted in Europe. In a study from Belgium, one town was subjected to an intensive media campaign on the avoidance of salt whilst the second control town was not advised in this manner (Staessen *et al.*, 1988). Despite a small but significant difference in urinary sodium excretion at the end of the study, there was no important effect on the average blood pressure in the low salt town. By contrast, a similar study from Portugal (a high salt consuming country) did demonstrate highly significant differences in blood pressure in the community randomized to receive salt restriction advice, compared with the control community (Fig. 9.3). (Forte *et al.*, 1989).

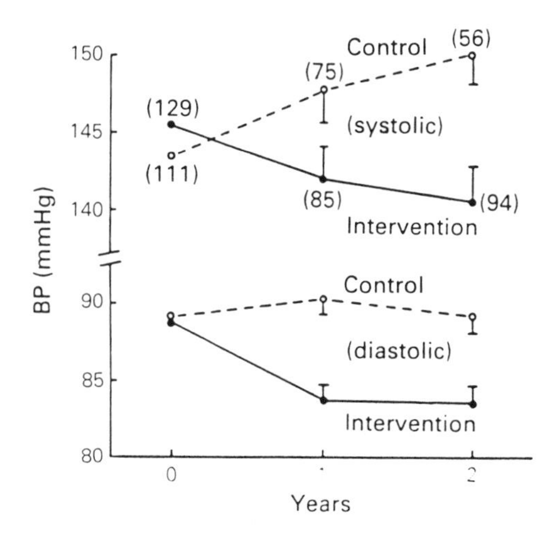

Figure 9.3 Effects on blood pressure of a population strategy to reduce salt intake, compared with a control community in Portugal (Forte *et al.*, 1989).

The salt hypothesis still generates a great deal of debate, but it is likely that no further reliable, large-scale epidemiological information will become available. However, additional well-conducted clinical studies in hypertensive patients are needed. It seems probable, however, that modest salt restriction from 150 down to 80 mmol sodium per day is associated with no harm to the individual and such dietary change can be achieved without great inconvenience. It simply means that the patient avoids notoriously salty foods and refrains from adding salt to food at the table. A diet which contains no salt whatsoever is totally unpalatable and is recommended by nobody. The original rice/fruit diet used by Kempner is no longer necessary as we now have potent antihypertensive drugs which have been shown to prevent cardiovascular diseases (particularly strokes) and prolong life.

Amongst patients receiving antihypertensive drugs, there is evidence, however, that salt depletion does have an additive effect. This is particularly so with the angiotensin converting enzyme inhibitors (MacGregor *et al.*, 1987). Many hypertensive patients are very pleased to make alterations to their diet or lifestyle in a manner that will help in the clinical management of their disease.

9.10 CONCLUSIONS

The salt hypothesis still remains controversial, but there is strong evidence in favour. Salt depletion does lower blood pressures in at least some hypertensive patients and there is also some evidence that lower salt diets might, on a national basis, lead to a reduction in the prevalence of hypertension associated with a smaller rise in blood pressure with advancing age. However, salt intake must be seen in the light of other dietary and environmental factors which affect blood pressure including obesity, high alcohol intake, potassium intake and intake of animal fat and dietary fibre.

REFERENCES

Ambard, L. and Beaujard, E. (1904) Causes de l'hypertension arterielle. *Arch. Gen. Med.*, **1**, 520–2.

Beevers, D.G. (1989) INTERSALT. *J. Hum. Hypertens.*, **3**, 279.

Dahl, L.K. (1961) The possible role of chronic salt consumption in the pathogenesis of essential hypertension. *Am. J. Cardiol.*, **8**, 571–5.

Elliott, P. (1989) The INTERSALT study: an addition to the evidence on salt and blood pressure and some implications. *J. Hum. Hypertens.*, **3**, 289–98.

Forte, J.T, Pereira Miguel, J.M., Pereira Miguel, M.J., *et al.* (1989) Salt and blood pressure: a community trial. *J. Hum. Hypertens.*, **3**, 179–84.

Huang Ti Nei Ching Su Wen. *The Yellow Emperor's Classic of Internal Medicine*, translated by I. Veith (1972) University of California Press.

INTERSALT Cooperative Research Group (1988) INTERSALT: an international study of electrolyte excretion and blood pressure. Results for 24 hour urinary sodium and potassium excretion. *Br. Med. J.*, **297**, 319–28.

Kempner, W. (1944) Treatment of kidney disease and hypertensive vascular disease with rice diet. *NC Med. J.*, **5**, 125–9.

Knudsen, K.D. and Dahl, L.K. (1966) Essential hypertension: inborn error of sodium metabolism. *Postgrad. Med. J.*, **42**, 148–52.

MacGregor, G.A. (1985) Sodium is more important than calcium in essential hypertension. *Hypertension*, **7**, 628–37.

MacGregor, G.A., Markandu, N.D., Best, F.E., *et al.* (1982) Double blind randomised crossover trial of moderate sodium restriction in essential hypertension. *Lancet*, **i**, 351–5.

MacGregor, G.A., Markandu, N.D., Singer, D.R.J., *et al.* (1987) Moderate sodium restriction with angiotensin converting enzyme inhibitor in essential hypertension: a double blind study. *Br. Med. J.*, **294**, 531–4.

Poulter, N.R., Khaw, K.T., Hopwood, B.E.C., *et al.* (1990) The Kenya Luo migration study: observations on the initiation of a rise in blood pressure. *Br. Med. J.*, **308**, 967–72.

Sever, P., Gordon, D., Peart, W.S. and Beighton, P. (1980) Blood pressure and its correlates in urban and tribal Africa, *Lancet*, **ii**, 60–4.

Staessen, J., Bulpitt, C.J., Fagard, R., *et al.* (1988) Salt intake and blood pressure in the general population: a community trial. *Hypertension*, **3**, 179–84.

Swales, J.D. (1988) Salt saga continued. Salt has only small importance in hypertension. *Brit. Med. J.*, **297**, 307–8.

Watt, G.C.M., Edwards, C., Hart, J.T., *et al.* (1983) Dietary sodium restriction for mild hypertension in general practice. *Br. Med. J.*, **289**, 432–6.

10 Electrolytes and blood pressure

D.A. McCarron and M.E. Reusser

10.1 INTRODUCTION

Although essential hypertension is clearly recognized as the most prevalent chronic disease in the industrialized world and a major contributor to the development of cardiovascular disease, stroke and renal failure, the optimal means of preventing and/or treating this disease remains an enigma. Pharmacological management is challenged in terms of its necessity in some cases and its effectiveness in others. While drug therapy is effective in preventing the complications of high blood pressure, its benefits must be weighed against the costs; pharmacological treatment for hypertension is life-long, expensive and associated with side effects. Therefore, non-pharmacological approaches, such as dietary interventions, hold a strong appeal as a means of managing hypertension. Although there is currently much controversy surrounding the effects of specific dietary factors on blood pressure, there is a rapidly expanding body of evidence to support a direct role of the diet in the development and treatment of hypertension.

10.2 HYPERTENSION AND DIET

The dietary factors that have been, and continue to be, examined most closely in terms of their effect on blood pressure are the electrolytes sodium, potassium, calcium and magnesium (McCarron, 1990; Luft, 1990). Dietary chloride, which was first suggested to play a role in blood pressure regulation more than 60 years ago (Berghoff and Geraci, 1929), is currently being reconsidered in the light of laboratory and human studies which indicate that, of the various sodium salts, only chloride increases blood pressure (Kurtz and Morris, 1983; Whitescarver *et al.*, 1984; Boegehold and Kotchen, 1989). In addition, epidemiological studies have identified a relationship between dietary phosphorus and the prevalence of hypertension.

Investigative efforts to define the specific roles of these nutrients in the development of hypertension have been extensive and varied. Several epidemiological studies have assessed the role of the intake of electrolytes in the development of hypertension (Harlan *et al.*, 1984; McCarron *et al.*, 1984; Reed *et al.*, 1985; Luft, 1990). Animal models have been studied to identify and characterize the mechanisms by which various electrolytes may influence blood pressure (Tobian *et al.*, 1985; McCarron, 1989; Luft, 1990). Clinical trials in normal and hypertensive individuals have examined the effect on blood pressure of modifying electrolyte intake (MacGregor, 1985; Luft *et al.*, 1989; Grobbee and Waal-Manning, 1990; Cutler and Brittain, 1990). However, despite vigorous efforts to resolve the questions surrounding the relationship between electrolytes and blood pressure, there remains strong disagreement among investigators supporting one view or the other, and general confusion for both health care providers and consumers.

Attempts to reconcile apparently conflicting or inconclusive data are confounded by the complexities of both the disease itself and the dietary factors that could potentially prevent or attenuate it. Essential hypertension is the end result of an interplay between multiple, interrelated regulatory systems, each of which is itself influenced by genetic variance and environmental factors. Nutrients are not consumed in isolation, but rather as interactive constituents of a total diet. Furthermore, dietary factors that influence blood pressure are not confined to electrolytes. There are numerous other aspects whose individual contributions may be as great as that of any single electrolyte. These include the amount of food consumed as reflected by obesity, the carbohydrate, protein and lipid content of the diet, the amount of alcohol consumed and the rate at which food is utilized as influenced by exercise. Recognition of the numerous possible combinations of these aspects in the total diet as well as the interactions among the nutrients themselves, together with the myriad of physiological, environmental and genetic factors influencing the development of hypertension, is essential to understanding the difficulties inherent in resolving the current controversies surrounding the role of the diet in blood pressure management. It may be that the differing views of skilled and committed scientists working in the area of nutrition are actually supportive concepts that, when properly integrated, will explain how the human diet contributes to blood pressure regulation (McCarron, 1990).

The practical aspects of these interrelationships have begun to be appreciated by the scientific and health care communities. A recent clinical trial of non-drug therapy included reduced intake of calories, salt and alcohol (Berglund *et al.*, 1989). The first prospective study to suggest that non-pharmacological approaches may be prophylactic actively concentrated on weight reduction, decreased intake of fat, alcohol and salt, increased potassium intake and aerobic exercise (Stamler *et al.*, 1989). The current physician education programme on the non-pharmacological management of hypertension of the National Kidney Foundation addresses each of these issues, in addition to several others, with emphasis on a tailored approach that combines individual features as well as general principles (National Kidney Foundation, 1989).

10.3 Salt

More than any other dietary component, salt has received the bulk of scientific attention. Although many of the long-held beliefs regarding the effect of salt on blood pressure have been discredited in recent years, they continue to be the general basis for dietary recommendations in the nutrition literature, the physician's office and national health policy (Surgeon General's Report on Nutrition and Health, 1988). Since nutrients express their physiological actions through integrated pathways, it is inconsistent to propose that lowering a population's intake of salt will result in uniform benefit to the entire population (Joint National Committee on Detection, Evaluation, and Treatment of High Blood Pressure, 1988). While some individuals may benefit, others will be unaffected or, conceivably, even adversely affected (Weinberger *et al.*, 1986; Miller *et al.*, 1987; Luft *et al.*, 1989). Grobbee and Hofman (1986) examined 13 prospective, randomized trials of salt restriction and identified significant reductions in blood pressure in only three of the studies. While a later study by MacGregor *et al.* (1989) reported more encouraging results, a larger and longer Australian intervention study produced mixed results (Australian National Health and Medical Research Council Dietary Salt Study Management Committee, 1989; Graudal and Galløe, 1989). Thus, demonstration that blood pressure response to moderate sodium restriction follows a Gaussian distribution is not surprising (Weinberger *et al.*, 1986; Miller *et al.*, 1987).

While it has been advocated that as a society we should reduce our dietary salt intake (Surgeon General's Report on Nutrition and Health, 1988), conservative application of this policy appears to be more appropriate for the population at large. It is estimated that about 70% of normal individuals and 50% of patients with essential hypertension are not 'salt sensitive', that is, negatively affected by salt intake in terms of blood pressure (Weinberger *et al.*, 1986; Miller *et al.*, 1987; Luft *et al.*, 1989). The challenge then is to be able to identify the individuals who are salt sensitive, which will be accomplished by the identification of the mechanisms that determine the heterogeneous blood pressure response to salt restriction. For example, one nutritional mechanism not related to electrolytes may be excessive caloric intake. Rocchini *et al.* (1989) were able to change a group of adolescent subjects from salt sensitive to salt resistant with a programme of weight reduction through dietary restraint and exercise. Thus, in these obese subjects, the rational management of their salt sensitivity was weight loss, not sodium restriction.

With elucidation of the mechanisms responsible for salt sensitivity, we may be able to explain why the relation between salt intake and blood pressure is not robust (Intersalt Cooperative Research Group, 1988), as well as why potassium intake when indexed against sodium intake is a predictor of blood pressure (Khaw and Barrett-Connor, 1984) and why the relation between the ratio of sodium to potassium and blood pressure is only evident at lower levels of dietary calcium intake (Gruchow *et al.*, 1988).

10.4 CALCIUM

More than 25 independent reports of epidemiological studies have identified a statistically significant association between the level of dietary calcium in the diet and either blood pressure or the risk of hypertension (McCarron, 1989). These reports have been based upon both prospective and retrospective evaluation of data from either well characterized regional populations or national data bases acquired by government organizations. The protective effect of adequate dietary calcium intake against the development of high blood pressure appears to be relatively independent of factors generally associated with hypertension, such as age, sex, race, physical activity, body mass index/weight, tobacco and alcohol consumption and other dietary components. Although this effect is independent of these other factors, several studies have indicated partial interactions among dietary calcium and several of these cofactors and blood pressure, perhaps the most important being dietary sodium.

Clinical trials directed at increasing dietary calcium intake indicate that a higher calcium intake lowers blood pressure in some patients with hypertension. Grobbee and Waal-Manning (1990) recently reviewed 22 such studies and found that a significant decrease in blood pressure was reported in nine. Several studies also report statistically significant blood pressure reductions in subjects with normal blood pressure when calcium is added to their daily regimen. The increase in dietary calcium to achieve these changes has been in the range of 500 to 1500 mg per day, well within the generally recommended allowance of dietary calcium. The conclusions of Witteman *et al.* (1989) from the Nurses Health Study are consistent with a blood pressure-lowering effect of a diet containing at least the recommended daily allowance (800–1000 mg) of calcium.

Calcium intake appears to be lower in hypertensive patients than in normal subjects, and the former also exhibit concurrent abnormalities in calcium homeostasis, particularly increased calciuresis (McCarron, 1989). In the report that first described the 'exaggerated' calciuresis of essential hypertension, it was noted that this metabolic aberration was more evident at higher levels of urinary sodium excretion (McCarron *et al.*, 1980). Similar and analogous findings have been reported by Kotchen *et al.* (1989) in the salt-sensitive Dahl rat, as well as by Kurtz and Morris (1985) in the hypertensive DOCA-salt rat. Kurtz and Morris (1983) postulated from their work in humans that the enhanced calciuresis may be a clue as to how sodium chloride raises blood pressure in salt-sensitive humans. Strazzullo and colleagues (1983) confirmed a higher urinary calcium excretion in hypertensive as compared to normal subjects. Luft *et al.* (1990) not only found relative hypercalciuria in hypertensive compared to normotensive individuals, but also showed that only sodium as the chloride salt augmented calcium excretion.

Findings from studies by McCarron *et al.* (1985) and Hamet and colleagues (1986) suggest that a reduction in dietary salt intake does not ameliorate the renal calcium excretory defect in a hypertensive animal model. Both groups observed that in the spontaneously hypertensive rat, an animal with well characterized disturbances of calcium metabolism that include a renal calcium leak (Young *et al.*, 1988), the antihypertensive effect of increasing dietary calcium intake required a normal or high-normal concurrent sodium chloride intake. Hamet *et al.* (1990) extended these observations to human subjects and recently presented similar data from a large cohort of hypertensive patients. Those individuals consuming higher levels of salt had either higher or lower blood pressure values depending on whether they were simultaneously consuming a diet low in calcium or a higher calcium diet that at least met the recommended daily allowance. Thus it is important to consider dietary calcium intake when suggesting modifications of other electrolytes in the diet. Recommendations to reduce consumption of specific foods to lower salt intake may result in simultaneous reductions in intake of major sources of dietary calcium, as well as potassium and magnesium.

10.5 POTASSIUM

Similarly, current efforts to reduce fat and/or caloric intake have focused on decreasing the consumption of dairy products, which results in reduced intake of not only the major dietary sources of calcium, but also those of potassium, 30–45% of which is derived from these foods. The proposal by Tobian (1986) that adequate dietary potassium afforded protection against hypertension to our ancestors, as it does to hypertensive animal models known to be salt sensitive, is supported by decades of research. In addition, increased potassium intake in the form of potassium chloride may lower blood pressure in hypertensive patients (Siani *et al.*, 1987), as well as reduce their medication requirements (Strazzullo *et al.*, 1990). The effect of potassium chloride appears to be engendered, at least in part, by the promotion of natriuresis. A reduction in medication requirements has been identified in hypertensive persons who increased their potassium intake by eating more vegetables (Strazzullo *et al.*, 1990). Krishna *et al.* (1989) demonstrated that short-term, severe potassium restriction promotes salt sensitivity in normal humans, as would have been predicted from the earlier epidemiological work of Khaw and Barrett-Connor (1987), who suggested that potassium intake protects against the potentially adverse effects of salt on arterial pressure.

The data for potassium supplementation as a suitable intervention are not sufficiently firm to allow guidelines for its general use. The overall safety of the intervention can be called into question in some patients with hypertension, particularly those with diabetes or those receiving drugs which impair the elimination of potassium or its deposition into cells. Increased potassium intake through an increase in consumption of fruits and vegetables appears the more prudent approach. Moreover, it is likely that such an increased potassium intake will reduce the incidence of vascular complications independent of effects on blood pressure (Khaw and Barrett-Connor, 1987).

10.6 MAGNESIUM

Simultaneous interactions of dietary sodium, potassium and calcium on the modulation of arterial pressure are probably not limited to those cations alone. In their analyses of data from the Honolulu Heart Study, Reed *et al.* (1985) were unable to isolate the beneficial effects of potassium and calcium from those of magnesium. Thus, dietary magnesium intake deserves further attention. Lower dietary magnesium intake was the other nutrient pattern identified in the initial report by McCarron *et al.* (1982) that described a link between low dietary calcium intake and high blood pressure. There are not yet sufficient data to define the aetiological role of magnesium deficiency in human hypertension or to support the use of magnesium supplementation as antihypertensive therapy in patients with essential hypertension (Altura and Altura, 1987; Reusser and McCarron, 1989). However, as with all nutrients, its adequate presence in the diet should be protected in recommendations of dietary modification.

10.7 OTHER ELECTROLYTES

In addition to sodium chloride, potassium, calcium and magnesium, other less commonly considered electrolytes may be important. In an analysis of nutrient intake in the HANES (Health and Nutrition Evaluation Survey) I data base, a potentially important effect of phosphorus was evident (McCarron *et al.*, 1984). Reed *et al.* (1985) noted a similar contribution of phosphorus intake in the Honolulu Heart Study data base. Research in laboratory models of hypertension supports the notion that a phosphorus supplemented diet promotes reduction in blood pressure (Bindels *et al.*, 1987). Although hypertensive patients commonly have serum phosphate values that are lower than normotensive control subjects, there are not as yet conclusive data from human investigations regarding the role of phosphorus in blood pressure control.

10.8 HORMONAL MECHANISMS

The hormonal mechanisms that modulate the effects that electrolytes exercise on blood pressure appear to function in an integrated fashion. For instance, the renin-angiotensin-aldosterone axis is pivotal in regulating sodium homeostasis and blood pressure. The manner in which this regulatory system behaves is greatly influenced by potassium homeostasis. A direct cardiovascular role for the calcium-regulating hormones has been suggested recently (Cross and Pang, 1980; Weishaar and Simpson, 1987). Studies by Resnick and colleagues (Resnick and Laragh, 1985; Resnick, 1990) in patients with primary disturbances in aldosterone secretion have demonstrated abnormalities in parathyroid hormone release. These investigators have also shown that manipulation of calcium intake and vitamin D in hypertensive patients results in simultaneous changes in renin levels which appear to move in an opposite and possibly coordinate fashion (Resnick *et al.*, 1985). Finally, evidence suggests that other humoral substances which have an independent effect on blood pressure may be elaborated by the parathyroid glands (Resnick *et al.*, 1990).

10.9 CONCLUSIONS

There is insufficient evidence to state unequivocally that too much dietary sodium (chloride) or too little potassium, calcium or magnesium is responsible for the genesis of essential hypertension, or that changes in the intake of these cations will consistently lower elevated blood pressure in patients with hypertension. However, in certain patients or subpopulations, alterations in electrolyte metabolism are clearly of importance. Although the mean change in blood pressure after altered cation intake in most studies is modest, individual changes may be significant. However, what little information is available regarding the reproducibility of changes reported in most studies is preliminary at best (Nowson and Morgan, 1989; Sharma *et al.*, 1989).

A sizable body of apparently conflicting data, and what appears to be disagreement among researchers, may actually represent the primary concepts that when properly aligned will clarify the influences of the human diet on blood pressure regulation. An important step in that direction is the recognition that nutrients are not consumed alone, but rather as interactive components of the total diet. The heterogeneous blood pressure response to manipulation of any individual nutrient may be one physiological manifestation of such dietary interactions.

As investigators, our current challenge is to identify the mechanisms that produce this heterogeneous response in order eventually to distinguish those patients and groups who would benefit from specific nutrient interventions and, in so doing, protect those who would not. As clinicians, our responsibility at the present time is to recognize and address both the direct and indirect effects of diet modifications. Responsible dietary recommendations should focus on the total diet, giving ample consideration to both the interrelationships of the nutrients themselves, as well as to what will be the aggregate effect of altering the intake of one nutrient on the intake of others.

ACKNOWLEDGEMENTS

The original work of Dr McCarron cited here, and preparation of this chapter, were supported in part by the National Dairy Promotion and Research Board.

REFERENCES

Altura, B.T. and Altura, B.M. (1987) Cardiovascular actions of magnesium: importance in etiology and treatment of high blood pressure. *Magnesium Bull.*, **9**, 6–21.

Australian National Health and Medical Research Council Dietary Salt Study Management Committee (1989) Fall in blood pressure with modest reduction in dietary salt intake in mild hypertension. *Lancet*, **i**, 399–402.

Berghoff, R.S. and Geraci, A.S. (1929) Influence of sodium chloride on blood pressure. *IMJ*, **56**, 395–7.

Berglund, A., Andersson, O.K., Berglund, G. and Fagerberg, B. (1989) Antihypertensive effect of diet compared with drug treatment in obese men with mild hypertension. *Br. Med. J.*, **299**, 480–5.

Bindels, R.J.M., van den Broek, L.A.M., Hillebrand, S.J.W. and Wokke, J.M.P. (1987) A high phosphate diet lowers blood pressure in spontaneously hypertensive rats. *Hypertension*, **9**, 96–102.

Boegehold, M.A. and Kotchen, T.A. (1989) Relative contributions of dietary Na and Cl to salt-sensitive hypertension. *Hypertension*, **14**, 579–83.

Cross, M.F. and Pang, P.K.T. (1980) Parathyroid hormone: a coronary artery vasodilatory. *Science*, **217**, 1087–9.

Cutler, J.A. and Brittain, E. (1990) Calcium and blood pressure: an epidemiologic perspective. *Am. J. Hypertens.*, **3**, 137S–146S.

Graudal, N. and Galløe, A. (1989) Effect of salt restriction on hypertension. Letter. *Lancet*, **ii**, 41–2.

Grobbee, D.E. and Hofman, A. (1986) Does sodium restriction lower blood pressure? *Br. Med. J.*, **193**, 27–9.

Grobbee, D.E. and Waal-Manning, H.J. (1990) The role of calcium supplementation in the treatment of hypertension – current evidence. *Drugs*, **39**, 7–18.

Gruchow, H.W., Sobocinski, K.A. and Barboriak, J.J. (1988) Calcium intake and the relationship of dietary sodium and potassium to blood pressure. *Am. J. Clin. Nutr.*, **48**, 1463–70.

Hamet, P., Daigneault-Gelinas, M., Lambert, J., *et al.* (1990) Epidemiological evidence of interaction between calcium and sodium intake impact on blood pressure: a Montreal study. *Proceedings of the 13th Scientific Meeting of the International Society of Hypertension*, Montreal, Quebec, Canada, July 1990, S124.

Hamet, P., Skuherska, R., Cherkaouil, L., *et al.* (1986) Calcium levels and platelet responsiveness in spontaneously hypertensive rats on high calcium diet. *J. Hypertens.*, **4**, (suppl. 6), S716.

Harlan, W.R., Hull, A.L., Schmouder, R.L., *et al.* (1984) Blood pressure and nutrition in adults. The National Health and Nutrition Examination Survey. *Am. J. Epidemiol.*, **120**, 17–28.

Intersalt Cooperative Research Group (1988) Intersalt: an international study of electrolyte excretion and blood pressure. Results for 24-hour urinary sodium and potassium excretion. *Br. Med. J.*, **297**, 319–28.

Joint National Committee on Detection, Evaluation, and Treatment of High Blood Pressure (1988) The 1988 report of the Joint National Committee on Detection, Evaluation, and Treatment of High Blood Pressure. *Arch. Intern. Med.*, **184**, 1023–38.

Khaw, K.-T. and Barrett-Connor, E. (1984) Dietary potassium and blood pressure in a population. *Am. J. Clin. Nutr.*, **39**, 963–8.

Khaw, K.-T. and Barrett-Connor, E. (1987) Dietary potassium and stroke-associated mortality: a 12-year prospective population study. *N. Engl. J. Med.*, **316**, 235–40.

Kotchen, T.A., Ott, C.E., Whitescarver, S.A., *et al.* (1989) Calcium and calcium regulating hormones in the 'prehypertensive' Dahl salt sensitive rat (calcium and salt sensitive hypertension). *Am. J. Hypertens.*, **2**, 747–53.

Krishna, G., Miller, E. and Kapoor, S. (1989) Increased blood pressure during potassium depletion in normotensive men. *N. Engl. J. Med.*, **320**, 1177–82.

Kurtz, T.W. and Morris, R.C. (1983) Dietary chloride as a determinant of 'sodium-dependent' hypertension. *Science*, **222**, 1139–41.

Kurtz, T.W. and Morris Jr, R.C. (1985) Dietary chloride as a determinant of disordered calcium metabolism in salt-dependent hypertension. *Life Sci.*, **36**, 921–9.

Luft, F.C. (1990) Electrolyte intake in the treatment of hypertension, in *Handbook of Experimental Pharmacology* (eds D. Ganten and P.J. Mulrow), Springer Verlag, Berlin, pp. 687–701.

Luft, F.C., Miller, J.Z., Lyle, R.M., *et al.* (1989) The effect of dietary interventions to reduce blood pressure in normal humans. *J. Am. Coll. Nutr.*, **8**, 495–503.

Luft, F.C., Zemel, M.B., Sowers, J.R., *et al.* (1990) Sodium bicarbonate and sodium chloride: effects on blood pressure and electrolyte homeostasis in normal and hypertensive mean. *J. Hypertens.*, **8**, 663–70.

McCarron, D.A. (1989) Calcium metabolism and hypertension. *Kidney Int.*, **35**, 717–36.

McCarron, D.A. (1991) A consensus approach to electrolytes and blood pressure: could we all be right? *Hypertension*, **17**, (suppl. I), 170–2.

McCarron, D.A., Lucas, P.A., Shneidman, R.J., *et al.* (1985) Blood pressure development of the spontaneously hypertensive rat after concurrent manipulations of dietary Ca^{2+} and Na^+: relation to intestinal Ca^{2+} fluxes. *J. Clin. Invest.*, **76**, 1147–54.

McCarron, D.A., Morris, C.D. and Cole, C. (1982) Dietary calcium in human hypertension. *Science*, **217**, 267–9.

McCarron, D.A., Morris, C.D., Henry, H.J. and Stanton, J.L. (1984) Blood pressure and nutrient intake in the United States. *Science*, **224**, 1392–8.

McCarron, D.A., Pingree, P.A., Rubin, R.J., *et al.* (1980) Enhanced parathyroid function in essential hypertension: a homeostatic response to a urinary calcium leak. *Hypertension*, **2**, 162–8.

MacGregor, C.A. (1985) Sodium is more important than calcium in essential hypertension. *Hypertension*, **7**, 628–37.

MacGregor, G.A., Markandu, N.D., Sagnella, G.A., *et al.* (1989) Double-blind study of three sodium intakes and long-term effects of sodium restriction in essential hypertension. *Lancet*, **ii**, 1244–7.

Miller, J.Z., Weinberger, M.H., Daugherty, S.A., *et al.* (1987) Heterogeneity of blood pressure response to dietary sodium restriction in normotensive adults. *J. Chron. Dis.*, **40**, 245–50.

National Kidney Foundation (1989) *Education Program on Nonpharmacologic Management of Hypertension*, NKF, New York, pp. 1–50.

Nowson, C. and Morgan, T. (1989) Effect of calcium carbonate on blood pressure in normotensive and hypertensive people. *Hypertension*, **13**, 630–9.

Reed, D., McGee, D., Yano, K. and Hankin, J. (1985) Diet, blood pressure, and multicollinearity. *Hypertension*, **7**, 405–10.

Resnick, L.M. (1990) Calciotropic hormones in human and experimental hypertension. *Am. J. Hypertens.*, **3**, 171S–178S.

Resnick, L.M. and Laragh, J.H. (1985) Calcium metabolism and parathyroid function in primary aldosteronism. *Am. J. Med.*, **78**, 385–90.

Resnick, L.M., Lewanczuk, R.Z., Laragh, J.H. and Pang, P.K. (1990) Plasma hypertensive factor (PHF) in essential hypertension (EH). *Proceedings of the 13th Scientific Meeting of the International Society of Hypertension*, Montreal, Quebec, Canada, July 1990, S101.

Reusser, M.E. and McCarron, D.A. (1989) Nondrug therapy of hypertension, in *Clinical Cardiovascular Therapeutics – Therapy of Hypertension*, (eds H.A. Punzi and W. Flamenbaum), Futura, New York, pp. 97–116.

Resnick, L.M., Muller, F.B. and Laragh, J.H. (1986) Calcium regulating hormones in essential hypertension. Relation to plasma renin activity and sodium metabolism. *Ann. Intern. Med.*, **105**, 649–54.

Rocchini, A.P., Key, J., Bondie, D., *et al.* (1989) The effect of weight loss on the sensitivity of blood pressure to sodium in obese adolescents. *N. Engl. J. Med.*, **321**, 580–5.

Sharma, A.M., Schattenfroh, S., Kribben, A. and Distler, A. (1989) Reliability of salt-sensitivity testing in normotensive subjects. *Klin. Wochenschr.*, **67**, 632–4.

Siani, A., Strazzullo, P., Russo, L., *et al.* (1987) Controlled trial of long-term oral potassium supplements in patients with mild hypertension. *Br. Med. J.*, **294**, 1453–6.

Stamler, R., Stamler, J., Gosch, F.C., *et al.* (1989) Primary prevention of hypertension by nutrition-hygienic means. Final report of a randomized, controlled trial. *JAMA*, **262**, 1801–7.

Strazzullo, P., Nunziata, V., Cirillo, M., *et al.* (1983) Abnormalities of calcium metabolism in essential hypertension. *Clin. Sci.*, **65**, 137–41.

Strazzullo, P., Siani, A., Giorgione, N., *et al.* (1990) Increasing dietary potassium intake reduces the need for antihypertensive medication. *Proceedings of the 13th Scientific Meeting of the International Society of Hypertension*, Montreal, Quebec, Canada, July 1990, S51.

Surgeon General's Report on Nutrition and Health (1988) US Department of Health and Human Services, Washington DC, pp. 1–78.

Tobian, L. (1986) High potassium diets markedly protect against stroke deaths and kidney disease in hypertensive rats – a possible legacy from prehistoric times. *Can. J. Physiol. Pharmacol.*, **64**, 840–8.

Tobian, L., Lange, J., Ulm, K., *et al.* (1985) Potassium reduces cerebral hemorrhage and death rate in hypertensive rats, even when blood pressure is not lowered. *Hypertension*, **7**, I110–I114.

Weinberger, M.H., Miller, J.Z., Luft, F.C., *et al.* (1986) Definitions and characteristics of sodium sensitivity and blood pressure resistance. *Hypertension*, **8**, (suppl. II), II127–II134.

Weishaar, R.E. and Simpson, R.U. (1987) Involvement of vitamin D_3 with cardiovascular function. II. Direct and indirect effects. *Am. J. Physiol.*, **253**, E675–E683.

Whitescarver, S.A., Ott, C.E. and Jackson, B.A. (1984) Salt-sensitive hypertension: contribution of chloride. *Science*, **223**, 1430–2.

Witteman, J.C.M., Willett, W.C., Stampfer, M.J., *et al.* (1989) A prospective study of nutritional factors and hypertension among U.S. women. *Circulation*, **80**, 1320–7.

Young, E.W., Bukoski, R.D. and McCarron, D.A. (1988) Calcium metabolism in essential hypertension. *Proc. Soc. Exp. Biol. Med.*, **187**, 123–41.

DISCUSSION

Jackson I did not know that blood pressure falls in summer, how much and why? I am quite impressed that Mediterraneans have a lower risk of blood pressure related disease. I wonder whether sunshine and vitamin D and calcium and blood pressure might not be all part of the same circle in some of the factors that we have been talking about. Could I ask whether in the United States of America the effect of calcium on blood pressure is a consideration that is used in deriving recommended daily amounts for calcium?

McCarron It has not been an issue. It has recently been concluded that it is premature to make blood pressure a basis for the RDAs. I might say that the blood pressure epidemiological data argue for an intake that is already within the range suggested by the National Academy of Science, and many people in the population, including males, are well under that 800 to 1000 mg/day level. This is not a linear relationship, it is an inverse sigmoidal relationship. There is a very steep portion of the curve and it appears that the break in the curve is around 600 mg/day in men and around 500 mg/day in women. This was evident in the Nurses Health Study and noted in the Honolulu Heart Study.

Jackson It may not be an issue in the United States, but in the United Kingdom, where the RDA for calcium is half what it is in the United States, it may in fact be something that the committees are considering.

Eastwood, Edinburgh Can I ask about the practical consequences of what you are saying? Yesterday, it was suggested that drinking milk was not a good idea because it increased lipid intake. One effective way of increasing a wide range of mineral intake, excluding sodium, is drinking milk. If one had the choice of increasing one's lipid with milk or dropping one's blood pressure by the calcium and other minerals in the milk, what should we be doing?

McCarron I think the first thing to do is to define what happens to lipids when you take a specific population and increase their dietary calcium intake for the purpose of lowering blood pressure and then carefully explore their lipid metabolism. In fact, we are doing this in Oregon; with Drs Roger Illingworth and Bill Connor, we are carefully assessing the changes in lipid metabolism during a doubling of dietary calcium intake through changes in dairy product consumption. As we have already reported at meetings, there has been no significant change in serum lipids. Different fats behave in different ways and just because there are saturated fats in dairy products, it may not be those that are going to have adverse impact upon lipid metabolism.

Kritchevsky, Philadelphia There are data suggesting that for reasons unknown, milk actually lowers cholesterol levels. In rats either skim or whole milk lowers cholesterol levels. There are some human studies that show that both milk and yoghurt lower cholesterol levels. I would like to ask Dr McCarron, in relation to the higher blood pressure amongst blacks in the United States, could that be related in some way to lactose intolerance?

McCarron Jim Sowers, who is head of the endocrine and hypertension group at Wayne State, has made this a major focus of his efforts for the last five years to try to tie together the sodium sensitivity of blacks to problems of lactose intolerance and low calcium intake. I know of three or four relatively well received publications that suggest that low calcium and low potassium intake may be part of the aetiology of hypertension in black subjects. We often forget that in the Nurses' Health Study and in the NHANES data, fluid milk, as the source of calcium, was also the biggest single source of potassium. So by taking care of one electrolyte, calcium, you take care of the other without even thinking about it.

Kritchevsky But actually, after listening to you and Dr Beevers saying data well received, it depends on the recipients.

Rosenberg, Boston There is one observation that interests me in regard to the elderly and that is the decline over the decades of life in the capacity to absorb calcium. This is one of the clear physiological changes which occurs with ageing, and the reciprocal observation about blood pressure. Has that relationship ever been looked at in longitudinal or population studies?

McCarron Not in terms of bio-availability or absorption. Bill Harlan has analysed the NHANES I and NHANES II data bases, and has concluded that there is a striking relationship between calcium intake with ageing and the rise in systolic blood pressure with age. This is an 'association' and not necessarily causation; vitamin D levels and PTH levels increase and intestinal absorption decreases after about age 50 or 55. There are some differences between men and women, but this is one of the generally agreed upon changes in metabolism that occur with the loss of the sex hormones. It would be interesting to see what happens to women who are prescribed oestrogens and oestrogen/progesterone combinations.

Beevers Vitamin D levels go down, I think?

McCarron While they initially increase and then they begin to fall, this differs between men and women. There is an initial peak after the age of 50 and then vitamin D production begins to decrease due to the kidney not producing normally.

Macdonald, London Why do we look at the cations? I wonder about the anions. The anion I have in mind, of course, is chloride because we know it is fairly important in the loop of Henle.

Beevers Some researchers have attempted to study this by administering sodium chloride or sodium bicarbonate and comparing the effects. Sodium bicarbonate loading did not affect the blood pressure as much as sodium chloride loading. The original hypothesis was that chloride raised blood pressure. When chloride intake was reduced, the sodium intake fell as a consequence as most of our salt is consumed in common salt. There was an attempt to reduce chloride intake, the sodium intake was a sort of bi-product.

McCarron The anion is important. You cannot induce a rise in blood pressure with any sodium salt except sodium chloride; on the other hand, you cannot induce a rise by just feeding chloride. It takes both the sodium and the chloride. I got into this whole issue in the mid-70s when I saw a problem of calcium handling by the kidney in a limited group of hypertensive patients. Renal calcium leakage is appearing as a very consistent marker of salt sensitivity. I think your observations hold potential to finally make salt sensitive hypertension a renal disease.

Fell, Glasgow I wonder if to make the confusion even worse I could ask Dr McCarron to say a few words about magnesium and phosphorus. In Glasgow, in conjunction with my consultant geriatric physician colleagues, we have been able to show very low levels of tissue magnesium and potassium in the elderly compared with younger populations, and dietary surveys also support the low magnesium intake. We are attempting to relate this to cardiac arhythmia symptoms.

McCarron I think the magnesium story is one that needs to be explored extensively. There is disturbingly little information about the relationship between magnesium intake and blood pressure. There is the old hard water/soft water story, although the principal constituent of hard water is calcium not magnesium. The Nurses Health Study suggests that dietary magnesium is almost as good a predictor as calcium. The two nutrients identified in the Nurses Health Study and in the report in *Circulation* were calcium and magnesium. If you summed the two over a 4-year period they accounted for a 35% reduction in new cases of high blood pressure after controlling for alcohol, exercise, smoking, weight, age, race, education, jobs, all the things that epidemiologists plug into the computer. As far as phosphorus is concerned, in the NHANES data there is a suggestion that people who are consuming balanced ratios of calcium and phosphorus are better off than those just getting higher intakes of phosphorus. It is easy to get a high intake of phosphorus in the Western diet without added calcium, but it is almost impossible to get a high level of calcium without an enhanced phosphorous intake because they are so closely linked. Dwayne Reed and his colleagues in the Honolulu Heart Study concluded that phosphorus is a modest predictor.

Hautvast, Wageningen Dr Beevers, there was a study in INTERSALT by the EEC in which more than a few thousand children were studied in Europe and the only relationship or association that was found was between blood pressure and magnesium, a weak association but the only one. I would like to stress that magnesium should be given more attention. Have you any data about magnesium?

Beevers No.

McCarron You cannot really use the INTERSALT data to discuss calcium and magnesium because the renal excretion of calcium and magnesium may have absolutely no relationship or bearing on intake or the way the body is handling it. I predict that INTERSALT will find a relationship between higher blood pressure and high levels of urinary calcium excretion. It has been suggested that this is because people eat too much calcium. The data around the world, including from the Netherlands, indicate that hypertensive persons do not consume too much. The problem is with the renal handling of calcium.

Blays, Bedford There has been a large number of studies quoted where individuals have been put on low sodium diets. I wondered how dietary compliance had been checked, because that is notoriously difficult to ascertain.

Beevers Compliance was usually checked by measuring urine sodium excretion. Most studies of sodium depletion in patients were very intensive projects in small numbers. There is a need for a trial of sodium restriction advice on a much larger scale. We did do this with alcohol advice, just to draw the parallel. We can get highly motivated people taking part in our alcohol lowering studies in the ward to obtain a fall in blood pressure. However, when we then did a similar study of alcohol advice on a larger scale, the results were very disappointing because compliance was poor. I think that compliance must be checked in all studies of dietary manipulation.

McCarron In Indiana, they had an NIH trial where they used professional people, presumably well motivated, easily trained individuals, and they gained compliance for about 8 to 12 weeks, after which urinary sodium started drifting back to baseline. People just do not stick with it. This is a problem. General advice as a strategy for the entire population has many problems. Short of mandating some absolute reduction in sodium content of the food source, I do not think it is feasible.

PART FIVE

Food Additives and Food Manufacture

11 Food additives: an overstated problem?

A.E. Bender

11.1 INTRODUCTION

A certain television advertisement started with 'Look, no Es'. The burden of the message was that the food advertised contained none of the dreaded chemicals designated by E numbers. The irony of this is that those so designated have been approved as being safe by the European Community.

Additives are used for a variety of functions: to preserve, to make the foods look and taste more attractive, to assist in their processing or for some specific benefit to the consumer. Examples include anti-spattering agents in fats anti-caking agents and antioxidants (Table 11.1).

11.2 USE OF FOOD ADDITIVES

By far the greatest number of additives are used for flavours – some 3500. About 99% (by weight) of the total amount of flavours used are extracts from foodstuffs. Other additives number about 250. The total amount used is about 148 000 tonnes a year, which is 0.37% of the total food consumed.

As a comparison, 200 compounds have been identified in honey, 400 in strawberries and several hundreds in a cup of tea. So 'natural', 'unprocessed' foods contain large numbers of strange chemicals, many of which have unknown effects in the body, and some of which are known toxins. The amounts, however, are very small and since man has been eating these foods for hundreds and thousands of years they cannot, in these quantities, be considered harmful.

11.3 TYPES OF FOOD ADDITIVE

Some additives are used to enable a particular product to be formulated, such as an emulsion (examples are, mayonnaise, chocolate, ice cream) or to

Table 11.1 Types of food additives (and numbers permitted)

1. Preservatives (14)	6. Flour 'improvers' plus processing aids (lubricants, anticaking agents, etc.)
2. Antioxidants (14)	7. Emulsifiers and stabilizers (56)
3. Colours (47)	8. Solvents (9)
4. Humectants	9. Sweeteners (12)
5. Flavours (3500)	10. Acidity controls (of which 85% are natural extracts)

Total weight per year (1985): 148 000 tonnes (=0.37% of total food tonnage)

facilitate the manufacturing process, to satisfy the demands of consumers or simply to please them. Another purpose is to assist consumers to carry out the advice of the nutritionist. For example, advice to eat less fat is largely ignored by the public, but low fat variants of preferred foods can be produced with the assistance of additives. For example, in order to reduce the fat in one low energy spread it became necessary to add three extra ingredients: casein to bind the extra water, improve mouth feel and enhance flavour release, gelatin to stabilize the product and extra salt for flavour and as a preservative, together with five additives: an emulsifier, a preservative, a colour (which might be considered unnecessary), an antioxidant and a flavouring ingredient.

11.4 SAFETY TESTING

Toxicology is not only an inexact science but it cannot provide a specific answer to the fundamental question: is it safe? Nothing is absolutely safe; indeed, an excess even of some vitamins can be lethal. Approximately 1 g sodium chloride is required daily; 100 g is approximately a fatal dose and on average 10% of a fatal dose is consumed daily. Many natural foods, ranging from broad beans to rhubarb, contain known toxic substances. Against this background, all the toxicologist can achieve regarding an additive is a reasonable assumption of safety when the substance is used under specified conditions.

There is no standard test to provide the required information, the techniques for testing for toxicity or for safety have been developing over the past 50 years.

The sequence usually followed is shown in Table 11.2. The chemical structure allows comparison with similar compounds. The acute toxicity test indicates whether the substance is worthy of further consideration. It helps to ascertain the target organ in the test animals as a first approach to attempting to determine the metabolic pathway of the substance.

The short-term toxicity test is a 90-day feeding trial with the test substance at several dose levels. There is always a dose level which damages tissues or

Table 11.2 Steps in safety assessment*

Chemical structure
Specification (purity)
Acute toxicity (target organ)
Metabolic handling
Short-term toxicity (90-day test)
Chronic toxicity (life-time tests/carcinogenicity)
Reproductive/teratogenicity studies
Mutagenicity, followed by likely human exposure, risk/benefit analysis (if necessary human studies)
In vitro tests

* Department of Health and Social Security (1982)

organs since most compounds are toxic when fed at a sufficiently high concentration, even ordinary and well accepted foods. The test substance has to be fed at very much higher concentrations than would ever be used as a food additive because no changes would be seen with the amounts (about 0.01–0.1%) used in foods.

In testing procedures dosage levels are related to body weight. Half the test animals are examined after 45 days and the remainder at the completion of the 90-day trial for any and every physiological, biochemical and behavioural change.

If at this stage the intended additive appears to be worthy of further testing, that is it has so far successfully passed appropriate tests, the next step is to examine for chronic toxicity, by feeding (obviously in amounts below that found to exert any ill-effects) for the life-time of test animals and through at least one reproductive cycle. Such tests seek carcinogenicity and embryotoxicity (including teratogenicity), that is whether there are any effects on the embryo, fetus or neonate. Two species of test animal are used, one of which should be other than a rodent.

In addition, a battery of tests for mutagenicity have been developed, both to examine the risk of causing mutations and to alert investigators to the possibility of carcinogenicity.

The likely exposure to the additive has to be taken into account; such as the required levels of use, in which types of food and the frequency of consumption of such food. The final judgment is passed to the legislators to assess risk-benefit. Since there is no such thing as *absolute* safety, legislators have to balance risk against advantage.

If the substance under trial is acceptable then the short-term tests provide information on levels that may be permitted in foods.

11.5 THE ACCEPTABLE DAILY INTAKE OF AN ADDITIVE (ADI)

The ADI is the amount that can be consumed on a daily basis for a life-time without appreciable health risks. From the short-term, 90-day trial the maximum no-effect dose is ascertained and this, divided by a safety factor of 100, provides the basis for the ADI. This is not a toxic threshold but provides an extremely large safety factor.

The factor of 100 is composed of two factors each of 10 – one to allow for differences between test animal species and man and the other to allow for biological variation in sensitivity to the compound.

It is worth comparing this with the levels of salt (and other natural substances such as caffeine) that we accept simply because they are looked upon as 'natural' and have a long, though not necessarily safe, record as part of the human diet.

11.6 CONSUMER PRESSURES, PRESS AND OTHER MEDIA DEMANDS

There is considerable confusion in the minds of the public in this very difficult area, not assisted by 'scaremongering' headlines. One result is that some bodies have advocated that unnecessary additives should be banned. They are. Additives can be used (according to the regulations) only if they are shown to be necessary for some specific purpose.

In assessing risk-benefit and the levels to be permitted, the legislator takes into account the dietary patterns of the population. This is one reason for differences in legislation between different countries; another reason may be differences of opinion.

11.7 GENERALLY REGARDED AS SAFE (GRAS)

Not many potential and other 'natural' additives have been through the exhaustive testing procedures described. Since each complete test takes 2 to 3 years to complete and is extremely expensive, it is not likely that they will be so tested – quite apart from constraints of testing facilities.

However, many compounds, such as metabolic intermediates, substances that have been safely consumed in some quantity for many years and substances that can be compared directly with others accepted as safe, can be considered under the classification 'generally regarded as safe' (GRAS). Examples include compounds which are included in the Krebs TCA cycle, those produced during the digestion of foods and those occurring naturally in foods, such as the fruit acids.

New substances such as artificial sweeteners or fat substitutes will require exhaustive testing to satisfy the authorities as well as consumers, if they are not already on the permitted lists.

Until some 50 years ago any substance could be added to foods provided that it was not listed as harmful. This system was easy to operate but not very safe for the consumer, since no one bore the responsibility for testing. The system has now changed to a permitted list, which is much more difficult to operate but requires the supplier or user to carry out tests to provide information for the legislating authorities.

Clearly, changing the system to a permitted list involved enormous numbers of investigations which took many years. Hence the change was effected in stages (the additives were dealt with in groups, such as colours, then preservatives, emulsifiers). Now all additives except the large list of flavours are on the permitted list system. Flavours are currently under investigation and since the great majority, as stated earlier, are extracts of natural materials, it is likely that most of them will be considered under the GRAS system. Otherwise, they will be banned until there is toxicological evidence of safety.

11.8 PROBLEMS

There are some fundamental problems. First, food manufacturers are trying to prove that a specific additive is not harmful, that is they are trying to prove a negative, which is not practicable.

Second, any possible test substance has to be fed at extremely high doses compared with those levels that will eventually be used in order to detect any effect – and at sufficiently high concentrations almost every possible dietary component can be shown to be toxic. It is rarely known whether there is a threshold below which there is no effect or whether the dose-response relation persists at low levels of feeding.

Man has a number of detoxicating mechanisms which enable the ingestion; without risk, of many harmful substances found in fresh and so-called 'natural' foods. Additives which fit into such mechanisms will be detoxicated in the same way but synthetic compounds, that is those that do not occur in nature, may not do so, hence the need, where possible, to determine the metabolic pathway of any proposed food additive.

The third problem is that the tests are carried out on animal 'models' and the results are extrapolated to human beings. Responses inevitably differ between species. Ascorbic acid's varying requirement between species demonstrates such anomalies.

11.9 FOOD INTOLERANCE

Quite a substantial proportion of the population reacts adversely to one or more foods, ingredients of foods or to an additive or to more than one of these combinations. This characteristic is a personal idiosyncrasy affecting about 1–2% of the population. Over 60 foods have been listed as causing allergy or some type of intolerance in some population groups. For them their only remedy is to avoid foods containing the offending substance – if they can identify it.

Generally it has been found that food intolerance is less widespread than people themselves believe. In other words, many who consider themselves to be 'allergic' are not so when tested.

About 150 people in every thousand react to 'something' in the environment; five to ten per thousand react to foods such as strawberries, potatoes, meat, fat or chocolate, while the figures for reactions to food additives are 0.3 to 1.5 per thousand according to the EC Science Committee for Food, or 0.1 to 2.6 according to the UK survey of 1984. Obviously with such small sample sizes it is difficult to be more precise.

A survey in 1980 on 1000 people was carried out with a response rate of 56.9%. Researchers believe that the vital questions asked about buying and eating habits were concealed. It was found that 1 to 4% of the respondents avoided specified foods for physiological reasons (such as headache, nausea, indigestion), as indicated in Table 11.3. The listed foods cover a wide range, from cereals and fruit to eggs and fish.

A data bank has been compiled jointly by the British Dietetic Association, Leatherhead Food RA, The Royal College of Physicians and the Food and Drink Manufacturers' Federation of foods known to be free from specific substances which commonly cause adverse reactions. Some of these

Table 11.3 Foods avoided because of adverse physiological reaction (from Bender, A.E. and Matthews, D.R. (1981) *Br. J. Nutr.*, **46**, 403–7)

	No. of subjects	*Percentage of sufferers*	*Percentage of total respondents*
Alcoholic beverages	36	19.4	6.4
Vegetables	30	16.2	5.4
Meat products	29	15.6	5.2
Cheese	27	14.6	4.8
Fish products	23	12.4	4.1
Other dairy products	13	7.0	2.3
Chocolate	14	7.6	2.5
Other sugar confectionery and sugar	8	4.3	1.4
Cereals	19	10.2	3.4
Fats and oils	10	5.4	1.8
Fruit	10	5.4	1.8
Eggs	5	2.7	0.9

Table 11.4 'Top ten' ingredients in Intolerance Databank*

Food specified 'free from':
Wheat and wheat derivatives
Milk and milk derivatives
Egg and derivatives
Soya
Cocoa
Antioxidants BHA and BHT†
Sulphur dioxide
Benzoate
Glutamate
Azo colours

* Intolerance Databank, Food and Drink Federation, 6 Catherine St, London, WC2B 5JJ.
† Butylated hydroxyanisole and butylated hydroxytoluene

substances are shown in Table 11.4 and include milk products, wheat and eggs. Among the additives are azo colours and the preservatives BHA, BHT, SO_2, benzoates and glutamate derivatives.

Any additive that can be omitted, even when it has been shown to be safe in use, might possibly benefit some individuals. In this respect but no other, those who have raised the invalid fuss about additives may have done some good.

11.10 SAFETY

It is true to say that apart from individual intolerance no additives have caused any harm in this country, and we must always add 'so far as we know'.

Some additives have been banned over the years but this has been the result of suspicion arising from animal experimentation and not from known harm to human beings. More recently, some additives have been banned in order to harmonize trade.

Indeed, in the world as a whole there has been only one intentional additive that, when used under legally permitted conditions (that is, apart from illegal overuse) has ever been shown to be harmful. This was the use of cobalt chloride to provide a good froth on beer, but which was shown in 1973 to be a cause of cardiac myopathy in Canada, Belgium and the USA, with some deaths.

Apart from this account, it is difficult to find any evidence of general harm – apart from intolerance – from any intentional additive. In fact, manufactured foods are often safer than unprocessed, so-called 'natural' foods, because no manufacturer would want to or be permitted to add substances such as cyanide to his products, whilst this and other toxins are naturally present in some foods. Indeed, there are several hundred deaths each year from puffer fish poisoning (from a toxin naturally present in the fish).

11.11 PUBLIC PERCEPTION

Consumers are faced with the difference between fact and fancy, between the public perception of risk as presented by a limited number of journalists and politically motivated and other publicity conscious individuals as opposed to authenticated facts. Understanding food labelling requires some knowledge of chemistry and physiology and there is, understandably, confusion among consumers about many matters in both the nutrition and food processing areas. This confusion is demonstrated by a survey in which the perceived list of so-called 'healthy' foods was headed by wholemeal bread, with sliced wholemeal bread lower down.

11.12 PUBLICISTS

The idea that additives are harmful has been headlined since some journalists learned about additives only when they were designated by E numbers. This, for a time, was supported by a faked list of harmful additives published on notepaper believed to have been stolen from Ville Juif Hospital in Paris. This myth started in 1976, spread later to other countries on mainland Europe, but came to the attention of British journalists much later. This list described many additives as harmful with one especially dangerous, that is **E 330**. This happens to be probably the most natural and harmless of all additives since it not only occurs in all fruits but is synthesized in the body – namely citric acid.

This type of misleading information has spread uninformed concern so that the term E number has become a symbol of 'unpleasantness' instead of safety, as was intended.

11.13 INDUSTRY'S RESPONSE

Right or wrong, consumer perception controls purchases, and food manufacturers have been obliged to respond to consumer perceptions about food additives. There have been several responses by the food industry. One is the use of 'natural' materials so that claims are made for foods 'free from artificial additives', or similarly 'free from colours, from preservatives'. This practice simply enhances consumer concern.

Another response has been to re-examine food formulations in an attempt to remove permitted food additives. This has been possible in some instances, partly because the food additives may not have been essential originally, because of a more rapid turnover of goods in supermarkets, because of advances in food technology or because the greater use of refrigeration has obviated the need for their use. A third approach has been to prepare modified and new versions of the food without additives so that, at the very least, consumers are offered a choice.

It is often said that colours are not essential but merely cosmetic. Earlier trials indicated that consumers preferred coloured products. Opinions may have changed more recently and similar but uncoloured foods are now available.

There seem to be more marketing opportunities for products labelled 'free from' then those with any positive virtues. One product (in the USA) claimed to be free from colours, preservatives, additives of all kinds, meat products, wheat products and soya – the claim was quite correct since the substance on offer was 'pure' inositol.

11.14 CONCLUSIONS

Much of the criticism of food is due to a reaction against technology. A few years ago technology was hailed as progress; for example, irradiation of food was lauded in the press as a major advance and it was the scientific world that tried to limit the plaudits. Now high technology is suspect in all walks of life; 'natural', home-made, what grandmother did, return to tradition – these are the key terms.

It is worth repeating that, apart from food intolerance, no harm has ever been laid at the door of additives in this country. When we consider how many toxic substances are present in natural, raw, unprocessed foods we can draw some comfort from the fact that manufactured foods are safer in this respect than many unprocessed ones.

REFERENCES

Bender, A.E. and Matthews, D.R. (1981) *Br. J. Nutr.*, **46**, 403–7.
Department of Health and Social Security (1982) *Guidelines for the Testing of Chemicals for Toxicity*, DHSS, Report on Health and Social Subjects No. 27, HM Stationery Office.
Joint report: food intolerance and food aversion (1984) *J. Roy. Coll. Phys.*, **18**, 83–123.
Reports of European Scientific Committee for Food (12th series) (1982), EUR, 7823.

12 The food manufacturers' role and problems

K.G. Anderson

12.1 INTRODUCTION

When I was asked to participate in this debate I was firstly flattered that such an audience might be interested in my contribution, and secondly somewhat alarmed because although my knowledge of nutrition might cover rather more than the back of the average postage stamp I felt fairly convinced that it could all be got down on the average postcard. When I saw the details of the programme I was comforted to see that the 'real' experts would be covering the complexities of cholesterol and fat, sugar, fibre and salt.

From the title of the conference I concluded that interest from the manufacturers' viewpoint should be directed at the current situation and what was likely to be happening in the relatively near future. I have also tried to consider some of the commercial difficulties faced by the food manufacturer in the free market of the West, which may perhaps not be readily apparent to nutritionists or dietitians.

Our very existence is about living with risk and about making judgments. It is of course possible to correlate eating and death with a 100% confidence level in either direction. If you do not eat you are certain to die, but if you do eat the certainty of death is just as absolute. The real differences are in the length of time death may be deferred and what its final cause might be. If you are an inhabitant of an area of the Third World stricken by famine and drought, the presence of a particular level of aflatoxin in an available food supply is of little consequence in the face of the immediate problem of short-term survival. This is an extreme case, but the balance between survival and indulgence is a relatively broad one, but with a moveable fulcrum depending upon the degree of risk seen to be acceptable at any given time and the state of knowledge relating to diet and health.

In case any of you have not had the chance to read it, I commend to you the British Medical Association publication *Living with Risk* which was the winner of the 1988 Science Book Prize and which is commendable for the perspective that it puts on these matters. That apart, it is a thoroughly good read, and the parts dealing with food and diet sensibly consider many issues of the day.

To try and get some perspective of my own I decided to start off by looking back, and in our small library in Croydon I was interested to see two books. First, *Nutritional Deficiencies in Modern Society*, the proceedings of a Food Education Society Symposium held in 1971, and second, *Nutrition and Lifestyles*, the proceedings of the First Annual Conference of The British Nutrition Foundation held in 1979. From this dip into the past it was interesting to realize that while two decades ago deficiencies were a primary cause of concern, by the turn of the last decade appreciation of the impact of choice and lifestyle on diet and health was becoming more predominant. Indeed, in the 1970s my own company first developed a policy of routinely screening newly developed products for their likely impact on the vitamin and trace element intake of consumers. Manufacturers have to face the situation that as their own techniques advance and their products become more sophisticated, so the reliance of susceptible groups on the nutritional integrity of those products increases.

More recently, greater attention has been focused on toxic hazards in foods and here we have an ever changing and advancing situation, where as hazards become recognized and identifiable through increasing sophistication of analytical and other techniques, so also do methods of control and elimination advance.

12.2 THE MANUFACTURERS' ROLE

Gorsuch (1989) has defined the role of the food industry as being:

> To deliver to the market place a choice of safe highly acceptable foods from which consumers can select a diet which is adequate to meet their nutritional, social, economic and organoleptic needs whilst at the same time maintaining the vigour and viability of the industry itself.

This proposition does relate primarily to the industry in the developed world and the consumers that it has to serve. I have included Gorsuch's paper in the references (it appeared in the Royal Society of Health collection of papers, *Diet and Health*) as it is not possible to consider it further in the time available.

In general, industrial policy will reflect Gorsuch's basic premise. That is to say, consumers should have freedom of choice and this should be exercised against a background of sound nutrition education and availability of nutrition information about products.

The industry has to supply the general public with the food that it demands in a democratic society where free market forces operate, but it is not for the industry to dictate to consumers what they should buy, rather the industry should make available a wide choice of products produced in accordance with the law and principles of good manufacturing practice, so that consumers can choose according to their individual preferences and requirements in selecting their diet. The industry needs to be responsive to changes in consumer demand and to scientific and technological advancement by marketing new or alternative products reflecting such changes and advances, and in providing appropriate information about them.

12.3 INDUSTRY'S ROLE IN PROVIDING INFORMATION

12.3.1 Labelling

Interested consumers should be well informed about the products they buy, particularly as to a food's ingredients, its energy value and nutrient content. This information requires to be provided within a legal framework to ensure that it is accurate, truthful and not misleading and at the appropriate level of detail. The commonly used means of providing this information is through labelling, but the useful role that can be played by advertising and other channels of communications should not be neglected.

The Food and Drink Federation (FDF), which is the 'umbrella' organization covering most of the trade bodies of the UK food and drink industry, encourages its members to provide nutrition labelling to assist consumers to choose how far they wish to follow nutritional guidelines. Labelling should be in a form that is truly informative, scientifically justified and not perjorative. Accordingly, the FDF has endorsed the Ministy of Agriculture, Fisheries and Food (MAFF) guidelines on voluntary nutrition labelling and has actively supported the establishment of compatible guidelines at the European Community level.

Nutritional claims should require the declaration of the claimed nutrients in accordance with the MAFF guidelines. Other health-related claims should be capable of being substantiated.

To reflect the evolutionary nature of nutritional science, the legislation applicable to nutrition labelling should be reviewed regularly.

12.3.2 Educational programmes

Educational programmes are essential to ensure that consumers can understand and use nutrition information in order to exercise choice between the products on offer and in the context of the messages received from a variety of sources.

Nutrition education has three phases: understanding the concepts, learning the terminology and acquiring the faculty of critical discernment. Such education requires that consumers be motivated to accept and make use of the information intended to influence dietary habits. Having received such education and information, it nevertheless remains for consumers to decide whether or not to modify their dietary habits. Accordingly, in planning education programmes there should be proper recognition of the need for consumers to accept responsibility for their own nutritional health and welfare.

Programmes should allow effective communication to specific categories of the population. All too often, consumers feel that generalized education and information do not apply in their individual situation. The programmes should therefore be specific, aimed at tackling genuine problems and relevant to the target group. To facilitate consumer understanding, such programmes should use appropriate terminology and ensure that educators are properly trained and equipped, and kept abreast of relevant developments in nutrition.

The responsibility for such programmes of consumer education does not lie primarily with the food industry. Industry can, however, be a useful partner in the education process by providing, as many companies already do, nutrition information and advice about products, including data for incorporation into national reference tables and data bases. As a contribution, FDF has produced a series of educational material under the title *Food Line*.

12.3.3 Nutrition research

Industry generally recognizes that nutritional science is under-resourced. The government bears the main responsibility for fundamental research on nutrition and related disciplines. The industry integrates the results of this work into its own research and development activities, some of which may indeed be fundamental, thus allowing the results of fundamental research to be applied to product design and food manufacturing.

The FDF has indicated that it supports the government's efforts to encourage:

1. dialogue at European level between the scientific and industrial communities, including those undertaking research programmes financed by the EC;
2. exchange of publicly available information in order to avoid duplication and to encourage cooperation;

3. coordination of methodologies so that the results of work (for example, epidemiological studies) are comparable.

The FDF fully recognizes the importance of bringing together the expertise and perceptions of the several bodies involved in, or concerned with, nutrition research.

12.4 MANUFACTURERS' PROBLEMS

The manufacturers' problems are of course manifold, and those which have an impact on nutrition may range from simple raw material supply problems and specific consumer requirements, to much more complex difficulties such as the provision of highly detailed nutrition information or the development and manufacture of products to meet specific or nutritional needs.

The problems can be broken down into six groups:

1. commercial viability;
2. the 'state of the art';
3. consumer perception;
4. the influence of philosophies;
5. specific needs;
6. communication.

12.4.1 Animal processing

Before discussing the problems in detail, I would like to give a brief overview of the complexities of animal processing. Part of the food industry in which I have worked was derived from the activities of that great scientist Baron Justus von Liebig. Visiting South America, Liebig was appalled at the waste of resource occurring as great herds of cattle were slaughtered purely for their hides and their carcasses left to rot at a time when parts of the world were short of meat. His observations stimulated his invention of the Liebig meat extraction process which was the origin of beef tea (*Extractum carnis*) and ultimately the Oxo cube. Development of the process eventually led to the curing of residual meat and the production of corned beef, together with the ultimate use of almost all parts of the carcass to produce everything from leather to the glue to stick the labels onto the cans. Some parts of the animals could be used for the production of microbiological culture media and pharmaceutical materials, this resulting in the establishment of Oxoid Ltd. In Fig. 12.1 I have shown some, but by no means all, of the complexity of the situation. With this in mind we may consider some of the groups of problems that were mentioned earlier. Commercial viability is obvious and needs no further elaboration.

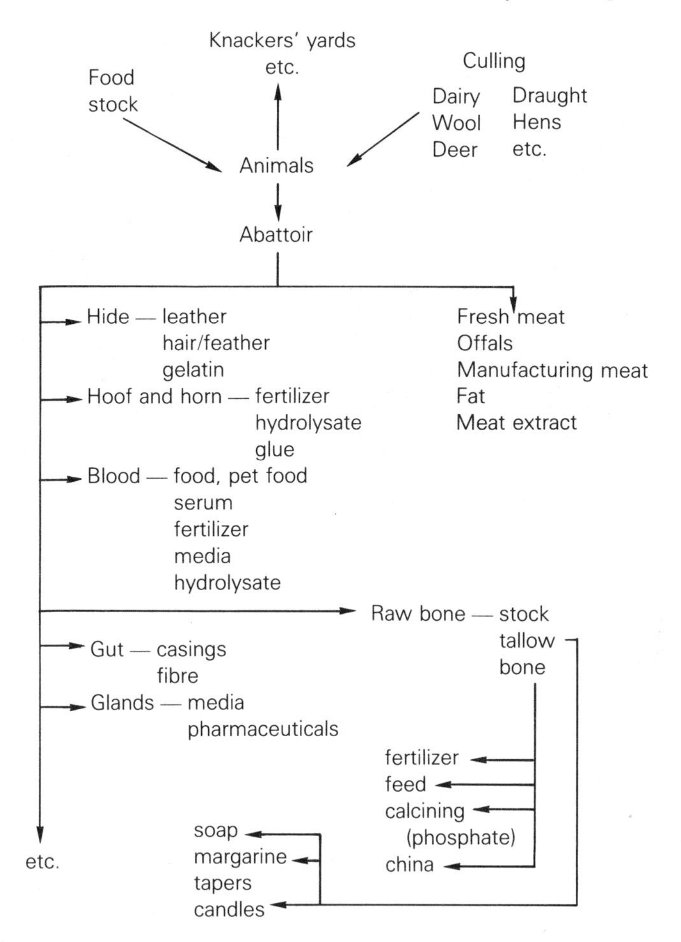

Figure 12.1 Animal processing

12.4.2 State of the art

Here the manufacturer needs to recognize that like good manufacturing practice, good nutrition is not a static concept and he needs to react and adjust to additional knowledge and changes of opinion.

12.4.3 Consumer perception

This is a difficult area for the manufacturer as it covers both consumer perception of what is required and consumer perception of the issues of the day. Confusion may be compounded by misinformation arising from a variety

Table 12.1 Topical issues

Additives
Aluminium
BSE
BST
COMA
Contaminants
Dioxins
Fat substitutes
Foreign bodies
Hormones
Irradiation
Microbiological safety
Microwave ovens
MRM
Organic foods
Sweeteners

BST = Bovine Somatotrophin
BSE = Bovine Spongiform Encephalopathy
MRM = Mechanically recovered meat
COMA = Committee on Medical Aspects of Health

of sources, such as media emphasis, public and political opinion, and by perceived risk and perceived benefit. I think, for example, of the perceived benefit of royal jelly. Table 12.1 shows a list of 'issues', some of which were recently identified by the Institute of Food Science and Technology Public Affairs Committee and the remainder I have added on my own account.

12.4.4 Philosophies

Here I simply mean philosophies to which consumers subscribe which make special demands on the manufacturer, for example vegetarian or vegan diets or particular religious or ethnic dietary requirements. In some cases, for example religious slaughter, conflicting demands are made on the manufacturer by different groups of consumers, leading to demands for even more complicated labelling. From a study of Fig. 12.1 it becomes evident that to maintain the balance and the established pattern of life there is a limit to the degree in which vegetarian diets can be supported without making significant changes of a quite wide-ranging nature, albeit that other developing technologies can replace animal-based industries in many cases. Of course, this is not to decry vegetarianism, but rather to draw attention to the potential effects that any constrained diet can have, even beyond the food chain itself.

12.4.5 Specific needs

Other specific needs must be recognized, for example of those who suffer from allergies or a particular dietary difficulty, such as the coeliac condition. There are also those who wish to follow a low cholesterol diet, or a low energy diet, and hence the development of 'light' products and fat substitutes such as Olestra and Simplesse.

Here it is appropriate to mention the Food Intolerance Databank set up by the industry with the Leatherhead Food Research Association. If this initiative is to survive and develop as it should, I recommend you to encourage its use wherever possible.

12.4.6 Communications

The Food Intolerance Databank provides a good lead in to the question of communications since it is designed to provide information to dietitians and other appropriate advisers. Communicating with consumers poses different problems, not least of which is that of meeting legislative requirements for labelling, and particularly nutritional labelling where back-up with significant analytical data is now required, at least in the UK. It is no longer good enough to work from published tables to calculate the data.

There are difficulties relating to space on labels, to consumer understanding, to consumer literacy and to confusion over the use of symbols. An alarming proportion of shoppers are not able to make use of the information provided on the pack as offered for sale on the supermarket shelf due to problems of literacy, of language, of visual defects and for other reasons. As we all know, nutrition information has been particularly difficult to convey, and has not really been satisfactorily resolved despite numerous efforts to come up with a simple and easily understood system.

12.5 THE FUTURE

What then of the future? I suspect that we shall really be looking at further development of existing trends, the extending influence of COMA, NACNE, etc.

Microbiological safety of food will continue to be an issue of the 1990s with Salmonella the major concern, although Listeria monocytogenes may take more of the limelight. Sugar and fat consumption will almost certainly continue to decrease and 'light' products increase market share significantly. We may expect lifestyles and social influences to continue to stimulate changes in the overall market and to see more 'designer products' aimed at specific groups such as the 'Greens', the elderly and people living alone, and to meet the needs of the 'browser' or 'grazer' with their more frequent and often mobile eating patterns. This will, of course, challenge us all to ensure that sound nutrition is not sacrificed, and on-going vigilance will certainly be needed by all involved.

12.6 CONCLUSIONS

In summary, the food and drink industry in the UK believes that proper nutrition should be achieved by well informed consumers exercising their freedom of choice in the market place. The industry recognizes that its principle role is to satisfy consumer requirements for safe foods at reasonable prices and to provide accurate information about the food supplies and their place in a healthy diet. Hence the need for consumer education and effective dialogue with consumer groups, opinion formers and other interested parties.

As convenor of the Institute of Food Science and Technology's Good Manufacturing Practices Panel, it would be remiss of me not to record the contribution that the IFST is making in this respect. I note particularly the publication, in 1986, of the booklet on the professional and scientific approach to the use of food additives and that published in 1989 on the nutritional enhancement of food. Nor should the role of the enforcement authorities in the UK and Europe and of the legislators be disregarded. Those who pass laws relating to food standards and labelling can help considerably to clarify the situation. On the other hand, they can also do much towards adding to consumer confusion if the ever increasing demands for information are not supported by ever improving education about what that extra information really means.

REFERENCES

Anderson, K.G. and Blanchfield, J.R. (eds) (1991) *Food and Drink – Good Manufacturing Practice: A Guide to its Responsible Management* 3rd 2nd edn, Institute of Food Science and Technology, London.

The British Market Research Bureau (1985) *Consumer Attitudes to and Understanding of Nutrition Labelling* Consumers Association/MAFF/National Consumers Council, London.

Conning, D.M. and Lansdown, A.B.G. (eds) (1983) *Toxic Hazards in Foods*, Croom Helm, London.

Dobbing, J. (ed.) (1988) *A Balanced Diet*, Springer-Verlag, Berlin.

IFST (1986) *Food Additives – the Professional and Scientific Approach*, Institute of Food Science and Technology, London.

Gorsuch, T. (1989) *Diet and Health*, Royal Society of Health, London.

Henderson, M. (ed.) *Living with Risk* (1987) The British Medical Association, John Wiley Chichester.
Howard, A.N. and Baird, I.McL. (eds) (1973) *Nutritional Deficiencies in Modern Society*, Newman Books, London.
IFST (1989) *Nutritional Enhancement of Food*, Institute of Food Science and Technology, London.
Turner, M. (ed.) (1980) *Nutrition and Lifestyles*, Applied Science.

DISCUSSION

Haustvast Professor Bender, I am not sure why we are unable to reassure the consumer that there is little need to worry about food additives. Why is the consumer so anxious about it? Why have we failed to communicate what you have told us?

Bender It is very simple. If you write a headline or an article in the newspapers saying the manufacturer is killing you, that you will all get cancer or you will die from microwave heating, it is published. If you wish to publish a scientific statement from this body saying 'this is not really true' it does not get published.

Hautvast Mr Anderson, do you think that the food industry is also suffering from the story about additives?

Anderson Yes, to a degree. Management in commerce today says with every problem there is an opportunity. Therefore if it appears that there is a problem because of microwave ovens, or the use of additives because of advanced processing, or the use of pesticides, then some manufacturers choose to use that as a way of selling their products, particularly if they are producing products that do not have additives or are not involved with pesticides.

Palmer, Cambridge Professor Bender started by mentioning an advertisement which said the product had no E numbers in it and yet neither lecture mentioned the biggest source of confusion about nutrition to the consumer, which is advertising. Most people get their nutritional information from consumer advertising. This is often extremely misleading, jumping on every new bandwagon, fear and fashion. The information is often inaccurate and, although I thought this was not allowed, makes claims for health. In fact, manufacturers on the Food and Drink Federation have been referred to the ASA for using misleading data. But it still goes on.

Bender The health claim is a serious and very difficult matter. The European Community said in the 1984 report that no health claims must be made. Two years ago the FDA in the States put out a discussion document that would allow health claims to be made. It is very confusing. Where do you draw the line? You must not say on the label 'vitamin A in my product is good for you' but you can put down an analysis of the product including the vitamin A. The other side of this is that anybody and everybody who has got a claim to make, does not make a statement for the negative side of their product. So when you advertise your orange juice as being rich in vitamin C, you do not say 'by the way there is no protein and not much zinc in this'. I have prepared a slide of a true nutritional labelling for an apple, which tells you how much waste, how much water and also puts next to the nutrients an opinion as to whether this is good or bad, you see high calcium is good, high sodium is bad. People want to know what does it mean. At the bottom I have put 'this is only my opinion anyway'. The practical problem is, manufacturers will make a claim, obviously, for anything they are selling, telling you only the virtues. Then you might ask, because you are a diabetic and you want to know about the sugar, or you are worried about osteoporosis and you think the calcium might help, what about the other things the manufacturer did not mention. This is the practical headache.

Anderson Yes I really support what Professor Bender has said. I do not approve of misleading advertisements. I would emphasize that there is a system to register complaints about misleading advertisements in this country.

Palmer Sorry, I have to come back. The system does not work, I'm afraid. I do not just mean misleading advertising. There is also the fact that no unprocessed food ever gets advertised. The consumer's information is biased towards the food industry. I have nothing against processed foods except that people do not recognize that porridge or oats are an alternative to the bran flakes or whatever. Most consumers get their nutritional information through vested interest sources whether it is on the label, the promotion, the packaging or the general distribution of products people can buy. If you live in an obscure rural area, you can only get certain foods. You do not have the choice. Also, when you say it is your role to give accurate information, it is, therefore, your duty to say for example that there is no protein in orange juice, or you can get this vitamin C from fresh fruit if you so desire. That would be the moral alternative.

Questioner unknown Can I ask both speakers to comment on what they think is going to be the likely use and the likely effects of fat substitutes and artificial sweeteners in manufactured foods in the 1990s?

Bender Artificial sweeteners, of course, we have had and accepted, but when it comes to what are called non-fat fats, we are going to see a very interesting situation. The first reaction, of course, is that this is terribly artificial, this is something manufacturers have made that we have never eaten before and we do not want it. The other side is if the nutritionist says that we ought to be eating a low fat diet, here is a method of tailor making a diet to suit what the nutritionist thinks is best. The biggest difficulty is that most of it is emotionally led, not scientifically led. In fact, when it comes to additives, some are banned in some countries but allowed in others and one of the deciding issues is whether or not the government pays attention to scientific fact or to emotional electoral appeal. We are seeing this now with irradiation. The British government has accepted the scientific aspect and has said we are going to lift the ban on irradiation. The German government has said we are bowing to the pressure of public emotional opinion.

Hautvast And this will be true of artificial fats?

Anderson Yes. Undoubtedly, the message has got through to the consumers or at least to a fairly large proportion of them about the need for fat reduction in the diet. It seems to me quite inevitable that if industry can use fat alternatives to produce products similar to those which previously contained a high level of fat, and they will be accepted by the consumer, then the industry will go right ahead and produce them.

Davidson, Dundee I would like to address a question to Mr Anderson. Some weeks ago I made a complaint to a manufacturer about a label which said 'all butter shortbread', but then underneath said 'no animal fats'. Now to me that was a rather ambiguous statement but the label remained on the supermarket shelves for some eight or nine months before the manufacturers finally agreed to remove it. Not because butter was an animal fat, but because so many other dietitians and nutritionists had complained about the label. Do you think this is the responsible labelling and the good practice that you say the FDF is supporting?

Bender No. I must say that it is the trading standards officer who should take care of that, complain to him.

Anderson You have to appreciate that those of us who work in the food industry or in any other industry or form of commerce live in a competitive society. Quite often the debate within one's own company may be of a nature that the marketing person comes with a planned activity and you say that you are not really very happy about that, and think it could be misleading. He then shows you that perhaps several competitors are making a similar description or a similar claim. The problem tends to be clarified through the enforcement authorities, the trading standards officers. In the UK there is a mechanism with LACOTS which is the Joint Consultative Committee on Trading Standards in cases such as the one you mention. If the person to whom you have made the complaint, and it is the trading standards officer's job to control this, is in doubt, he can seek a view from LACOTS. If they support the point that has been made, then they will advise all the local authorities throughout the UK of their dissatisfaction with that product label and thus indirectly persuade the company to remove it from the market. If it does not it may be prosecuted.

Garrow, London I was very intrigued by Mr Anderson's second point in the manufacturers' problem. He referred to perceived benefit and quoted royal jelly. I wondered if he would enlarge on that a little bit, previously he had been talking about Oxo, which was the thing that the Tommy wanted to have sent out. I believe that there is no benefit from royal jelly and not much from Oxo, does the industry have any responsibility to transmit this information – you may take it because you like it, but not because it will confer any health benefit.

Anderson That is a difficult one. I can say to you that I commented on royal jelly purely because I happened to have seen a part of a consumer television programme, where the amount of royal jelly as a food component was ridiculously small. Even if it had had any of the properties that are attributed to it they were most unlikely to have done any good.

Cannon, London I enjoyed Professor Bender's use of the term idiosyncratic reaction, not his invention of course. I look forward to United Biscuits suggesting that some individuals have an 'idiosyncratic reaction' to saturated fat. Indeed, I think Tate and Lyle have already suggested that some individuals are 'intolerant' to sugar. More seriously, I would agree with Professor Bender that the issue of toxicity of additives has been grossly overblown. The issue is not additives as contaminants so much as additives as adulterants – a form of legalized fraud. Take the British sausage, for example. If you look at the label of a British sausage you will you see well towards the end the number 128, which may or may not by now have an E number, this is a red dye. Question: what is a red dye doing in a sausage? Answer: it makes the fat look like meat. This is one of countless examples. Or take flavours: somebody from the trade has written that if flavours were banned or at least severely limited, half the manufactured food made in this country could not be marketed – it would disappear. There is nothing wrong with manufactured food as such. The problem is the load of saturated fat and processed sugar in the British food supply. The fact is that if additives, especially cosmetic additives, were used more responsibly by the manufacturers, the volume of fat and sugar in the British food supply would decrease.

Lean, Glasgow I was interested in Mr Anderson's link with the drug industry at the end of this talk. Going back to his earlier comment about responsibility for nutrition research being with the government, I wonder if he can comment at all on the sources of research and development funding, and the disparity between the food industry and the drug industry in the amount they put into research. Very broadly, do you think there is any likelihood of movement or change in that sort of disparity?

Anderson I am not sufficiently familiar with the differences between the pharmaceutical industry and the food industry to comment. Historically much of the UK funding of food research has been by the government and particularly through MAFF and the load that the industry has borne directly has been relatively light. I suppose that it has increased a bit over the last couple of decades. Certainly when I was at Leatherhead Food Research Association, the funding was something like 2:1, industry to government. That ratio has changed, industry contributes more now but curiously there is more direct funding of fundamental work from government. Funding from government sources, as you know only too well, has diminished. I am not saying that it has increased from industry to meet the shortfall. I cannot make a comparison with the pharmaceutical industry, I just do not know enough about it.

Bender It is obviously very much bigger, but we live in a real world and everybody works for profit. There is more profit in pharmaceuticals than there is in food, therefore the pharmaceutical industry puts a lot more money into research. There are two and a half thousand PhDs in chemistry in this country and twelve hundred of them work for ICI. Obviously the great majority would be doing research, and an enormous amount of effort goes into seeking new drugs, because one single good drug can make or break a company. Many of our big companies are profitable because of a small number of successful drugs, so that the incentive to do research is obviously very much greater than the food industry because there is no one food, unless you start inventing bread today, in which you can make a very big profit.

Moynahan, London Sydenstricke, who was familiar with the pellagra in the hillybillies in the United States, considered the British rationing during World War II and was surprised that the population was not suffering from some form of vitamin B deficiency. One of the reasons suggested was the role of condiments, these do not usually appear in nutritional surveys and studies. Bovril and Oxo were produced by the processing of food to enable transportation across the Atlantic. Bovril and Oxo contain essential nutrients. Apropos the whole discussion, we must remember that in animal husbandry the function of a salt lick is to provide micro-nutrients that may not be otherwise obtained from the diet. Man may need the same systems as condiments.

McCarron, Oregon As I sit here I am struck once again by an attitude and perspective that either the food industry is trying to poison society for the purposes of commercial ends, or that we eat as a social factor, losing sight of the fact that we have to eat to live. Essential nutrients are 'essential' because without them normal physiology cannot occur. The human organism and all living species have a remarkable capacity to deal with excesses. They have absolutely no means of dealing with deficiencies. When there is a deficiency of an essential vitamin or mineral the body has to make a trade-off which has consequences. The kidney eliminates excesses of potentially lethal potassium and sodium in the course of the day. I think we need to recognize that many of the problems we are dealing with are still diseases of interactions and deficiencies, and less likely to be due to poisoning foods. We need to take a more optimistic and more intellectual view and try to understand on a scientific basis.

PART SIX

Nutritional Recommendations

13 Nutritional policy for children

B.A. Wharton

13.1 INTRODUCTION

Since 1974 three editions of *Present Day Practice in Infant Feeding* have been published by the Department of Health (Oppe *et al.*, 1974, 1980a, 1988). These have been followed by welcome changes and improvements in infant feeding, but there are at least three areas of uncertainty where policy is unclear, our advice inconsistent, and practice variable.

13.2 SUCCESSES

13.2.1 Recommendations

The reports made broadly the same recommendations with differing degrees of emphasis and, possibly not surprisingly, it is the 1974 recommendations that have been the most important. The first recommendation was that even in a developed country such as Britain there were still a number of advantages in babies being breast fed. That was a major recommendation. It does not sound very strange now, but in 1974 it needed to be said quite loudly. The second one was that if babies were not breast fed then they should not receive cows' milk. Almost all of the bottle feeds in 1974 were in effect whole cows' milk modified only by the addition of vitamins and iron. The third recommendation was that the introduction of solid foods ('weaning') should be later than it had been. Some mothers were introducing solid foods into the diet of their babies at even two weeks of age. Amongst the wealthier more adventurous the food liquidizer encouraged this trend, but it was taken up by all socio-economic groups.

13.2.2 Changes in practice

Those three recommendations were made and, quite surprisingly, there followed distinct changes in feeding practices. It is not clear why these reports

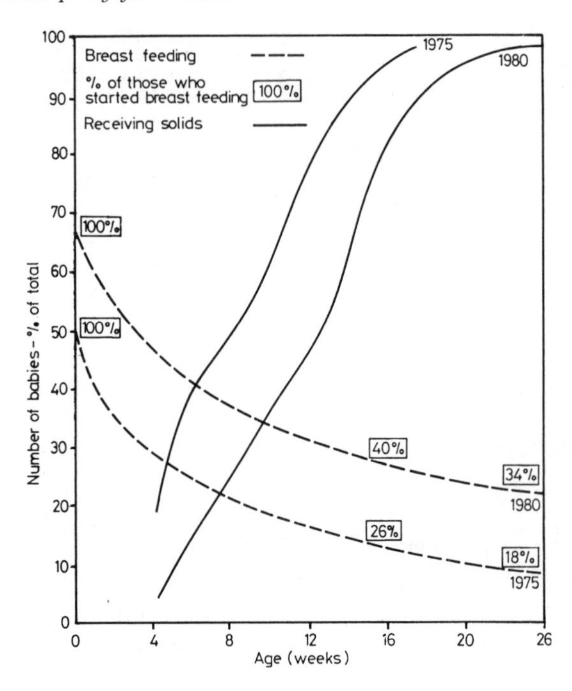

Figure 13.1 Proportion of babies in England and Wales breast feeding and receiving solid foods during the first 6 months of life in 1975 and 1980. (From Wharton, 1982a, with permission of the editors of *Arch. Dis. Child.*)

were so successful because few people take much notice of recommendations made by government committees and the general population changes its behaviour very slowly. But there were changes and these are summarized in Fig. 13.1 (Wharton, 1982a). In 1975 about 50% of newborn babies were put to the breast but subsequently there was a rapid fall off so that by 8 weeks less than 20% were still being breast fed. Five years later, two-thirds of the babies were initially breast fed and although there was again a fall off, about one-third of them were still breast fed at 8 weeks. The age of introduction of solids in 1980 was about 8 weeks later than in 1975. By 1980 almost all bottle-fed babies were receiving an infant formula whose composition met the nationally agreed guidelines (Oppe *et al.* 1980b). The practice of feeding whole cows' milk to infants before 6 months of age had almost completely disappeared.

13.2.3 Changes in child health

Mostly it is difficult to show that policy statements are followed by changes in eating behaviour. It is also difficult to show that the changed eating behaviour

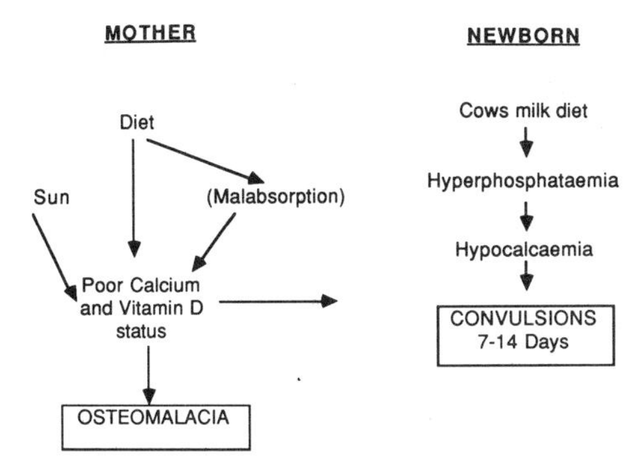

Figure 13.2 Maternal osteomalacia and neonatal hypocalcaemic convulsions.

is followed by an improvement in health. However, not only were the recommendations followed by changes in infant feeding practices, they were also followed by improvements in child health.

(a) *Neonatal hypocalcaemia*

In the 1970s about 2% of babies convulsed, because they were not being fed properly (Fig. 13.2). They were receiving whole cows' milk and hence a high phosphate load. This syndrome has almost completely disappeared. If it is seen now, knowing that the postnatal diet is satisfactory, that is it is either breast milk or a modern infant formula (both of which have a low phosphate content), attention is focused on the antenatal diet of the fetus. Now in many instances the occurrence of postnatal hypocalcaemia in a baby is an indication that the mother's vitamin D status is poor (Fig. 13.2) and there is the satisfaction of treating her and making her healthier for her next pregnancy.

(b) *Gastroenteritis*

Regarding gastroenteritis, Britain was the 'sick man of Europe' in 1972 (Table 13.1). The infant mortality rate due to gastroenteritis was six times the rate occurring in some other European countries. The high cardiovascular death rates in the UK have attracted considerable attention, but they demonstrate nothing like the international gradient for gastroenteritis detailed in Table 13.1. By 1983, however, this position had changed so that mortality figures are comparable with those of many other European countries.

Is this substantial reduction in gastroenteritis mortality (much greater than has occurred in infant mortality from other causes) due to changes in infant feeding practice? The evidence is circumstantial and based on analysis of

Table 13.1 Infant mortality from gastroenteritis in selected EEC countries (rates per 100 000 live births)

	1972	*1983*
Netherlands	9	0
West Germany	20	3
France	25	6
Belgium	35	5
England and Wales	55	6

sequential trends. Oral rehydration therapy, the only other possible cause of this rapid reduction in mortality rates, was not used in this country very much in the 1970s. The plausible explanation is that more babies are breast fed (as indicated in Fig. 13.1), hence there is less gastroenteritis. Therefore fewer babies are admitted to hospital with this condition. The babies who are admitted are less metabolically ill; the prevalence of hypernatraemia, a particular complication of dehydration in babies fed cows' milk, has decreased dramatically from about 40% of those admitted in the 1970s to only about

Table 13.2 Prevalence of hypernatraemia (plasma sodium >150 mmol/l) (from Wharton, 1982b)

Year	*Glasgow number of cases (and deaths)*	*Newcastle number of cases (and as % of all with gastroenteritis)*	*Sheffield number of deaths mainly sudden (and per 1000 live births)*	*London number of cases (and as % of all with gastroenteritis)*
1971		27(28%)		55(18%)
1972	22(3)	22(21%)		12(4%)
1973	28(4)	20(17%)	10(1.6)	23(5%)
1974	22(1)	2(2.5%)	5(0.7)	
1975	9(2)	7(12.5%)	1(0.2)	
1976	9(4)		1(0.2)	11(3%)
1977	3(0)		0(0)	5(1%)
1978	3(0)		0(0)	
Reference	Arneil and Chin (1979)	Pullan *et al.* (1977)	Sunderland and Emery (1979)	1971–2 Tripp *et al.* (1977) 1973–7 Manuel and Walker-Smith (1980)

5% (Table 13.2). Thus fewer babies develop the illness and as those that do are less ill, they are easier to look after and fewer of them die.

(c) Coeliac disease

The later introduction of solid foods possibly linked with other changes in infant feeding practices, such as more breast feeding and more highly modified bottle feeds, has been followed by a considerable reduction in coeliac disease in the childhood population (Table 13.3).

13.2.4 Perspective

So there have been successes in infant feeding practices with recommendations followed by changes in the behaviour and measurable improvements in child health. This demonstrates a success in central policy and illustrates the role that central government can play. It is, of course, recognized that confirmation of such data is required before trying to modify the diet of whole populations.

13.3 PROBLEMS AND UNCERTAINTIES

Against this background of success, are there any difficulties, areas where policy is uncertain, either because it has been overlooked or it is not known

Table 13.3 Prevalence of coeliac disease (from Wharton, 1982b)

Study		*Prevalence per year*						
		1973	*1974*	*1975*	*1976*	*1977*	*1978*	*1979*
Leeds (Littlewood *et al.*, 1980)	Cases per year (expressed as a 3-year average)	11	9	6	4	2	<1	<1
Taunton (Challacombe and Bayliss, 1980)	Cases per year		11	4	7	2	1	1
Glasgow (Dossetor *et al.*, 1981)	Cases under 2 years per 10 000 live births	3.5	2.5	4	1	1	0	
Malmö (Sweden) (Lindberg, 1981)	Cases per 1000 live births	1	1	1.1	1	1.1	1.0	(1)

what it should be? Three areas are discussed: vitamin supplements, the 'milk' part of the diet of older infants and fat intakes of toddlers.

13.3.1 Vitamin supplements

Rickets still occurs, although less than previously. The conventional wisdom, particularly in the last few years, is that vitamin D status is less concerned with diet (Poskitt *et al.*, 1979). Perhaps there should not even be a recommended daily amount for the vitamin because vitamin status depends mainly on an individual's exposure to sunshine. Nevertheless, Britain's population lives in a northern country much further north, for example, than the USA. These islands would be uninhabitable or they would be like Greenland if it were not for the Gulf Stream. Scotland in particular is a northern country with limited amounts of sunshine and so there is concern that the contribution of the diet should be ignored completely in relation to vitamin D status.

This concern is underlined by observations in Birmingham (Fig. 13.3). Asian toddlers who were not receiving vitamin supplements had low plasma concentrations of 25-OH vitamin D in the winter, showing the expected rise to more satisfactory levels in the summer months (although not to the extent seen in white children in a neighbouring town). Those children who were receiving dietary supplements, however, maintained satisfactory levels throughout the year.

Policy changes relating to vitamins have caused considerable confusion. The statement made in the 1980 edition of *Present Day Practice in Infant Feeding* is likely to be preferred because it is logical and allows professional discretion. It recommends that all children should have vitamin drops unless the professional adviser is confident that children have an adequate supply from elsewhere in the diet. Sadly this recommendation was not popular amongst the majority of health care professionals because it was not didactic enough. The present statement is a blanket one that all children should have vitamin drops from 6 months onwards. This statement is interesting since it does not mention recommendations before the age of 6 months, although it is known that breast milk contains very little vitamin D.

13.3.2 Milk part of the diet in older infants

Figure 13.4 shows the haemoglobin distribution in children attending a well baby clinic in Birmingham. A considerable number of them have a haemoglobin of less than 11 g/dl, that is clinical anaemia according to the WHO definition. The prevalence of anaemia was 18% in white British

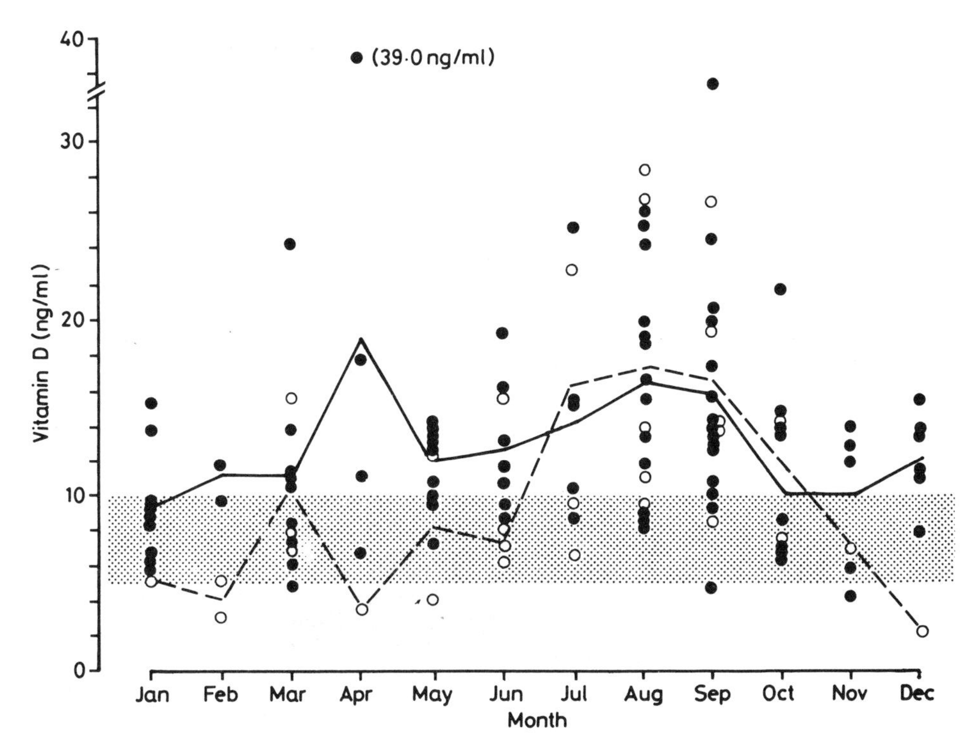

Figure 13.3 Plasma vitamin D (mean and individual values) in Asian toddlers according to seasonal variation and vitamin supplementation. ●—● = Vitamin-supplemented children; ○---○ = unsupplemented children. Plasma vitamin D <10 ng/ml = suboptimal (shaded area); <5 ng/ml = deficient. (From Grindulis *et al.*, 1986, with permission of the editors of *Arch. Dis. Child.*)

children and 27% in Asian toddlers. There have now been various surveys throughout the country in various cities with broadly similar results. Anaemia is common; this seems to be due to simple iron deficiency, probably of dietary origin. Perhaps mild anaemia causes little harm, but there is increasing evidence from a large number of studies that it is associated with, and at least partly results in, psychomotor delay. Some of these studies are summarized in Tables 13.4 and 13.5.

The author's view originally was that children living in a poor inner city environment did not get much stimulation which led to psychomotor delay; nor did they receive such a good diet which in turn, led to iron deficiency; thus there is associated cause and effect. However, the intervention studies summarized in Table 13.5 using double-blind, controlled randomized techniques demonstrate a cause and effect between iron deficiency and psychomotor delay. At least it is fairly easy to diagnose iron deficiency and

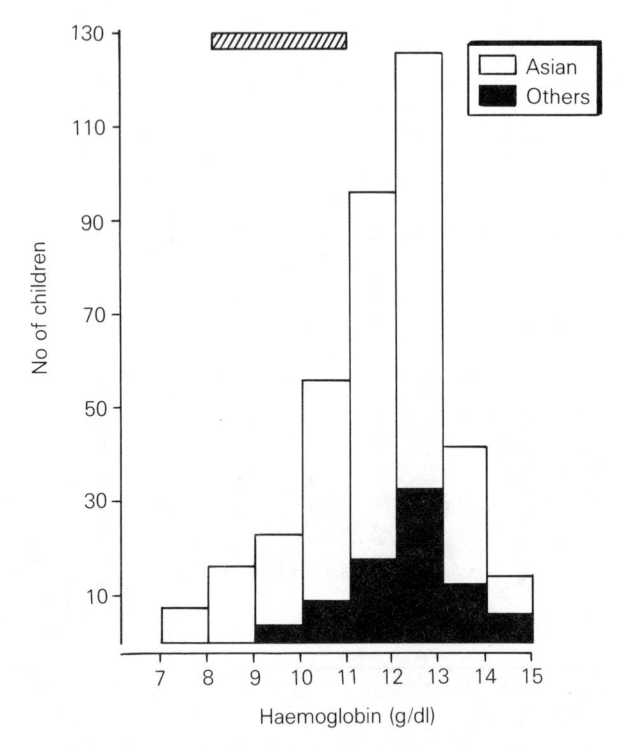

Figure 13.4 Distribution of haemoglobin concentration in 470 children aged 17–19 months attending four child health clinics in inner Birmingham. (From Aukett *et al.*, 1986, with permission of the editors of *Arch. Dis. Child.*)

treatment costs very little on a daily basis, whereas defining inner city deprivation is not straightforward and treatment is unclear.

Prevention of iron deficiency is therefore clearly important. There is ample evidence that continued use of an iron and vitamin C fortified infant formula (or a follow-on milk) is an effective way of achieving prevention (Olivares *et al.*, 1989).

13.3.3 The 'toddler gap'

Nutritional policy for adults has been dominated by fat, fibre, salt and sugar. These are summarized in the COMA report (DHSS, 1984) and are applied to individuals from the age of 5. The infant feeding reports go up to the age of 1 mainly, and there are a few recommendations for older children, such as that they should all receive vitamin supplements and that mothers might begin

Table 13.4 Association of iron deficiency anaemia and iron deficiency with 'poorer' psychomotor development (from Parks and Wharton, 1989)

	Lozoff et al., 1982	*Walter et al., 1983*	*Lozoff et al., 1987*	*Grindulis et al., 1986*		*Deinard et al., 1981*	*Deinard, et al., 1986*	*Oski et al., 1983*
Age group (months)	6–24	15	12–23	21–23		18–60	11–13	9–12
Iron status	Deficient Anaemic	Deficient Anaemic	Deficient Anaemic	Deficient Anaemic		Deficient Anaemic	Deficient Non-anaemic	Deficient Non-anaemic
Haemoglobin (g/dl)	<10.5	8.5–11	<10	<11		<11	HCT * >34	>11
Number	28	10	34	55		25	34	18
Controls								
Haemoglobin (g/dl)	>12	>11	>12	>11		>11	Sufficient HCT >34	Sufficient >11
Number	40	12	35	79		21	157	10
Developmental scale	Bayley MDI	Bayley MDI	Bayley MDI	Sheridan Fine motor	Social	Bayley MDI	Bayley MDI	Bayley MDI
Study group	86.6	98	96.6	20.6	20.6	92	109	84.6
Control group	100.4	113	104.6	21.5	21.4	98	109	93.7
	$p < 0.0025$	$p < 0.0025$	$p < 0.05$	$p < 0.05$	$p < 0.01$	NS	NS	NS

* HCT = haematocrit

Table 13.5 Effects of prolonged Iron therapy on development (from Parks and Wharton, 1989)

	Aukett et al., 1986	*Lozoff et al., 1987*	
Age group (months)	17–19		12–23
Iron status	Iron deficient Anaemic	Iron deficient Anaemic	Iron deficient Non anaemic
Haemoglobin (g/dl)	8–11	<10.5	>12
Number	48	52	21
Controls	Iron deficient Anaemic	Iron sufficient Non anaemic	Iron sufficient Non anaemic
Haemoglobin (g/dl)	8–11	>12	>12
	49	36	36
Treatment	Oral iron and vitamin C or vitamin C	IM iron/oral iron/ placebo	IM iron/oral iron/placebo
Duration (months)	2	3	3
Response	42% effectively treated achieved 6 + new skills vs 13% of controls (p<0.02)	Effectively treated achieved 10 + increase in score, controls did not (p<0.007)	No significant change/in score (NS)

to consider giving semi-skimmed milk to their children over 2. Therefore, between 2 and 5 years, and possibly betwen 1 and 5 there appears to be a 'gap' where recommendations are sparse and professionals give differing advice. Perhaps there is insufficient information to make recommendations, but certainly problems occur because of this gap. The occasional mother, perhaps more health conscious than average, gives toddlers a diet containing semi-skimmed milk, and large amounts of fibre from cereals and vegetables. The fat content and hence the energy density of the resulting diet is low; children just cannot eat sufficient to meet energy requirements and growth falters.

Ways of closing this 'toddler gap' in nutrition policy should be examined during the next year or so.

13.4 PERSONAL PRACTICE AND RECOMMENDATIONS

While national policies and 'consensus statements' are awaited, what should be advised? The intakes of saturated fat, iron and vitamins are very much

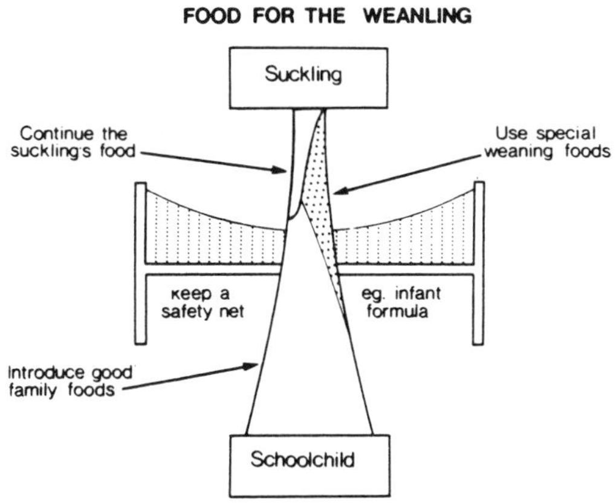

FOOD FOR THE WEANLING

Suckling

Continue the suckling's food

Use special weaning foods

Keep a safety net

eg. infant formula

Introduce good family foods

Schoolchild

Figure 13.5 The weanling's bridge from suckling to schoolchild with a 'safety net' of fortified foods. (From Wharton, 1986, with permission of the editors of *Acta. Paediatr. Scand.*)

affected by the milks or infant formulae consumed. Therefore, if correct policies are achieved for milks and infant formulae then more than half the battle is won. The author has recently given his own views concerning milks for babies and children (Wharton, 1990). These are summarized in Tables 13.6 and 13.7.

The concept developed previously for the older infant and the young toddler is shown in Fig. 13.5. Children cross a dietary bridge from days as sucklings 'mewling and puking' as Shakespeare said, to the relative safety of the diet of the schoolchild. Initially they will continue with the suckling's food, either breast milk or an infant formula. They will start to receive special weaning foods, either home-made or bought commercially. Eventually they will receive healthy adult foods so that by the age of 5 the COMA recommendations should apply (DHSS, 1984). But many of them fall off this bridge into vitamin D deficiency or iron deficiency and a few of them experience growth faltering due to protein energy deficiency. For such children some sort of safety net is needed. In the UK's particular circumstances, that safety net should be an infant formula or a follow-on milk because their use will ensure an adequate intake of vitamin D and iron. In general, I am against the principle of treating whole populations and this is what fortification means. The history of fortifying foods and so treating whole populations is not a good one in this country. Consider, for example, the epidemic of hypercalcaemia (Wharton and Darke, 1982). Mothers taking the advice of the nutritional and medical establishment poisoned their children with vitamin D.

Table 13.6 Available milks and infant formulae, analysis per 100 ml (and per 100 kcal)

	Energy (kcal)	*Protein (g)*	*Vitamin D (μg)*	*Iron (mg)*	*Saturated fat (g)*	*Sodium (mmol)*
Breast milk	70	1.3 (1.9)	0.01	0.08	2.1 (3.0)	0.6 (0.9)
Infant formulae (1)	67–70	1.5–1.9 (2.5)	1.0 (1.5)	0.4–0.7 (0.8)	1.0–1.9 (2.1)	0.6–1.1 (1.2)
Follow-on milks (2)	65–67	2.0–2.9 (3.7)	1.1–1.2 (1.6)	0.7–1.2 (1.4)	1.2 (1.8)	1.3–1.5 (2.1)
Cows' milk						
Ordinary	67	3.4 (5.1)	0.02*	0.05	2.5 (3.7)	2.2 (3.3)
Semi-skimmed	48	3.4 (7.1)	0.02*	0.05	1.1 (2.3)	2.2 (4.6)
Skimmed	34	3.4 (10.3)	0.02*	0.05	—	2.2 (6.7)

* Some varieties contain added vitamin D: see note 3

† Infant formulae and follow-on milks reconstituted from powder; liquid cows' milk; the Cow and Gate and Wyeth products are also available as liquids ready to feed at 16–20 p per 100 ml

1. Whey-based formulae: Aptamil (Milupa), Ostermilk (Farley's), Premium (Cow and Gate), SMA Gold Cap (Wyeth)
   Casein-predominant formulae: Milumil (Milupa), Ostermilk 2 (Farley's), Plus (Cow and Gate) SMA White Cap (Wyeth)
2. Junior Milk (Farley's), Progress (Wyeth)
3. Some ordinary, semi-skimmed, and skimmed milk UHT tetrapaks have added vitamin A (approx. 130 μg) and D (approx 0.5 μg per 100 ml) but such products are not widely available and this level of fortification is very much lower than in infant formulae or follow-on milks

Table 13.7 Use of milks and infant formulae (numbers in parentheses refer to the footnotes)

Age	*Milk part of diet*	*Need for vitamin D supplements (5)*	*Other dietary comments*
0–6 months	Breast milk	Variable. When in doubt give them (6).	When received in adequate volumes breast milk or an infant formula will provide the nutrient requirements of normal babies (8). Very few will require solid weaning foods before 3 months, but almost all will want something extra by 6 months.
	or Infant formula (1)	No (6)	
6–12 months	Breast milk	Yes (7)	As part of a mixed diet. Care is necessary to ensure an adequate intake of absorbable iron.
	or Infant formula (2) containing added iron	No (7)	As part of a mixed diet. For choice between infant formula or follow-on milk – see note (9)
	or Follow-on milk (3)	No (7)	As part of a mixed diet (9).
	not Ordinary cows' milk (4)	—	
12–24 months	Ordinary cows' milk	Recommended (7)	As part of a mixed diet.
	or Follow-on milk (3)	No (7)	This is not current British habit but in some countries follow-on milks are used in the toddler years and may have advantages (10).

Table 13.7 Continued

Age	*Milk part of diet*	*Need for vitamin D supplements (5)*	*Other dietary comments*
2–5 years	Semi-skimmed milk	Recommended (7)	To support normal growth an adequate supply of energy from other dietary sources is necessary. There is little point in limiting the saturated fat intake from milk if this is replaced by other sources of saturated fat, e.g. butter.
	or Cows' milk	Recommended (7)	
	or Follow-on milk (3)	No (7)	
5 years	Skimmed, semi-skimmed or ordinary cows' milk		As part of a diet in which fat provides no more than 35% of energy (11).

1. A whey-based formula is preferable (see Table 13.6). Casein-predominant formulae are acceptable. There is no objective evidence that casein-predominant formulae are more satisfying than whey-based ones, but switching is common and although unnecessary is probably harmless.
2. All of the infant formulae listed in Table 13.6 have added iron and they all conform to the national guidelines given in *Artificial Feeds for the Young Infant* (Oppe *et al.*, 1980b).
3. Junior Milk (Farley's) or Progress (Wyeth) (see Table 13.6).
4. This author does not recommend ordinary cows' milk before the age of 1 year. It contains little vitamin D and iron and causes subclinical but significant gastrointestinal bleeding in about one-third of children (Ziegler *et al.*, 1990). Other possible disadvantages are its higher concentrations of saturated fat and sodium, but the significance of this at this age or for the child's future is not clear. The extra cost of using an infant formula or a follow-on milk rather than ordinary cows' milk (10–15 p per day) is small compared to the price of other baby products.
5. The recommended dose of supplementary vitamin D is 7 μg daily. This is provided by 5 daily Department of Health (DH) vitamin drops (not prescribable on NHS prescription) or by many proprietary preparations. Vitamin policies have changed a number of times over the years and there are many different views, hence the detailed explanations in the column.

6. Supplements are not formally recommended by the DH for children aged less than 6 months. Ideally the mother should have received vitamin D supplements in pregnancy, but very few do. If there is any doubt about the mother's vitamin D status during pregnancy (e.g. Asian mothers, winter pregnancies, northern parts of the country) then her breast-fed baby should be given a vitamin D supplement. Bottle-fed babies receive sufficient vitamin D from infant formula.
7. Vitamin drops are formally recommended by the DH for all children aged over 6 months (Oppe *et al.*, 1988). Few receive a supplement, yet few develop rickets – presumably because the input of vitamin D from the sun and the diet are adequate. However the, plasma concentration of 25-OH vitamin D is very low in many children.

   Special efforts should be made to ensure that the following groups receive vitamin D supplements or are drinking a milk containing added vitamin D.

   (a) Breast-fed babies over 6 months of age – they are growing rapidly so requirements are high, stores from birth will be reducing, breast milk contains little vitamin D; sunshine might not provide enough.
   (b) Children having only limited exposure to the sun, such as those living in northern, urban areas, not having a sunny holiday, Asian children or for other cultural, social or medical reasons.
   (c) Children on vegetarian diets.

Despite the blanket recommendation by the DH that all children aged over 6 months receive a supplement they are not essential for children who are drinking a reasonable amount (say 500 ml) of an infant formula or a follow-on milk, since all these products contain adequate amounts of vitamin D. However, if a child received an infant formula or follow-on milk plus a vitamin D supplement of 7 μg the total dose would not be toxic.

8. Successful breast feeding has advantages over bottle feeding even with modern infant formulae and even in Britain. In Dundee (Howie *et al.*, 1990) babies breast fed for 13 weeks or more (but not less) had, when compared to bottle-fed babies, fewer gastrointestinal upsets, even after allowing for various 'confounding' factors.
9. There is little to choose between an infant formula and a follow-on milk at this age. Theoretically, the lower protein/energy ratios in infant formulae would not adequately support a mixed diet which was very low in protein, for example fruits and 'sweets' only. In practice and in careful studies, this does not seem to be a problem, but if there is any doubt then follow-on formula should be used. Some mothers wish to move on from an infant formula seeing it as a sign of welcome development by their babies – here a follow-on milk, *not* ordinary cows' milk is-recommended.
10. The arguments for extended use of a follow-on formula into the toddler years are that they contain added iron and vitamin D and have limited contents of saturated fat and sodium, without any limitation of energy content. Its use enables a mother to follow a healthy eating policy for her family without any risk of compromising the energy intake of her growing toddler. It is only part of the diet, however, and many other foods eaten by toddlers, particularly 'fast foods', could result in very high intakes of saturated fat and sodium. The follow-on formulae in Britain all have reduced saturated fat content, but this will not be compulsory if the draft EC directive becomes law (Astier-Dumas *et al.*, 1983); some European follow-on formulae contain mostly cows' milk fat.
11. The general recommendations of the COMA report on diet and cardiovascular disease (DHSS, 1984) apply from 5 years of age. The report's general recommendations on fat, fibre, salt, etc., specifically exclude the under-5s.

Nevertheless, the epidemiological evidence for iron deficiency and its possible effects seem to indicate that an intervention at population level is appropriate.

REFERENCES

Arneil, G.C. and Chin, K.C. (1979) Lower solute milks and reduction of hypernatraemia in young Glasgow infants. *Lancet*, **ii**, 840.

Astier-Dumas, M., Fernandes, J., Marquardt, P., *et al.* (1983) First report of the Scientific Committee for food on the essential requirements of infant formulae and follow up milks based on cows milk proteins, in Food – Science and Techniques 1983; *Report of the Scientific Committee for Food*, 14th Series, EUR 8752 EN, Commission of the European Communities, Luxemburg, pp. 9–32.

Aukett, M.A., Parks, Y.A., Scott, P.H. and Wharton, B.A. (1986) Treatment with iron increases weight gain and psychomotor development. *Arch. Dis. Child.*, **61**, 849–57.

Challacombe, D.N. and Baylis, J.M. (1980) Childhood coeliac disease is disappearing. *Lancet*, **ii**, 1360.

Deinard, A., Gilbert, A., Dodds, M. and Egeland, B. (1981) Iron deficiency and behavioural deficits. *Pediatrics*, **68**, 828–33.

Deinard, A.S., List, A., Lindgren, B., *et al.* (1986) Cognitive deficits in iron deficient anaemic children. *J. Pediatr.*, **108**, 681–9.

DHSS (1984) Diet and cardiovascular disease. *Report on Health Service Subjects No. 28*, HM Stationary Office, London.

Dossetor, J.F.B., Gibson, A.A.M. and McNeish, A.S. (1981) Childhood coeliac disease is disappearing. *Lancet*, **i**, 322.

Grindulis, H., Scott, P.H., Belton, N.R. and Wharton, B.A. (1986) Combined deficiency of iron and vitamin D in Asian toddlers. *Arch. Dis. Child.*, **108**, 681–9.

Howie, P.W., Forsyth, J.S., Ogston, S.A., *et al.* (1990) Protective effect of breast feeding against infection. *Br. Med. J.*, **400**, 11–16.

Lindberg, T. (1981) Coeliac disease and infant feeding practices. *Lancet*, **i**, 449.

Littlewood, J.M., Grollick, A.J. and Richards, I.D.G. (1980) Childhood coeliac disease is disappearing. *Lancet*, **ii**, 1359.

Lozoff, B., Brittenham, G.M., Viteri, F.E., *et al.* (1982) The effects of short term oral iron therapy on developmental deficits in iron deficient anaemic infants. *J. Pediatr.*, **100**, 351–7.

Lozoff, B., Brittenham, G.M., Wolf, A.W., *et al.* (1987) Iron deficiency anaemia and iron therapy. Effects on infant development test performance. *Pediatrics*, **79**, 981–95.

Manuel, P.D. and Walker-Smith, J.A. (1980) Decline of hypernatraemia as a problem in gastroenteritis. *Arch. Dis. Child.*, **55**, 124.

Olivares, M., Walter, E., Hertrampf, F., *et al.* (1989) Prevention of iron deficiency by milk fortification. *Acta. Paediatr. Scand. [Suppl.]*, **361**, 109–13.

Oppe, T.E., Arneil, G.C., Baum, J.D., *et al.* (1988) Present day practice in infant feeding: third report. *DHSS Report on Health and Social Subjects*, **32**, 1–66.

Oppe, T.E., Arneil, G.C., Creery, R.D.G., *et al.* (1974) Present day practice in infant feeding. *Report on Health and Social Subjects No. 9*, HM Stationary Office, London.

Oppe, T.E., Arneil, G.C., Davies, D.P., *et al.* (1980a) present day practice in infant feeding – 1980. *Report on Health and Social Subjects No. 20*, HM Stationary Office, London.

Oppe, T.E., Barltrop, D., Belton, N.R., *et al.* (1980b) Artificial feeds for the young infant. *DHSS Report on Health and Social Subjects No. 18*, HM Stationary Office, London.

Oski, F.A., Honig, A.S., Helm, B. and Howanitz, P. (1983) Effect of iron therapy on behaviour performance in nonanaemic iron deficient infants. *Pediatrics*, **71**, 877–80.

Parks, Y.A. and Wharton, B.A. (1989) Iron deficiency and the brain. *Acta Paediatr. Scand. [Suppl.]*, **361**, 71–7.

Poskitt, E.M.C., Cole, T.J. and Lawson, D.E.M. (1979) Diet, Sunlight, and 25-hydroxy vitamin D in healthy children and adults. *Br. Med. J.*, **1**, 221–3.

Pullan, C.R., Dellagrammatikos, H. and Steiner, H. (1977) Survey of gastroenteritis in children admitted to hospital in Newcastle-upon-Tyne 1971–5. *Br. Med. J.*, **1**, 619.

Sunderland, R. and Emery, J.L. (1979) Apparent disappearance of hypernatraemic dehydration from infant deaths in Sheffield. *Br. Med. J.*, **3**, 575.

Tripp, J.H., Wilmers, M.J. and Wharton, B.A. (1977) Gastroenteritis: a continuing problem of child health in Britain. *Lancet*, **ii**, 233.

Walter, T., Kovalskys, Stekel, A. (1983) Effect of mild iron deficiency on infant mental development scores. J. Pediatr., **102**, 519–22.

Wharton, B.A. (1982a) A quinquennium in infant feeding. *Arch. Dis. Child.*, **296**, 32–7.

Wharton B.A. (1982b) Past achievements and future priorities – a view of present-day practice in infant feeding 1990, in *Nutrition and Health, a Perspective* (ed. M.R. Turner), MTP Press, Lancaster, pp. 169–81.

Wharton, B.A. (1986) Food for the weanling – the next priority in infant nutrition. *Acta. Paediatr. Scand. [Suppl.]*, **323**, 93–8.

Wharton, B.A. (1990) Milk for babies and children. *Br. Med. J.*, **301**, 774–5.

Wharton, B.A. and Darke, S.J. (1982) Infantile hypercalcaemia, in *Adverse Effects of Foods* (eds E.F.P. Jelliffe and D.B. Jelliffe), pp. 397–404.

Ziegler, E.E., Fomon, S.J., Nelson, S.E., *et al.* (1990) Cow milk feeding in infancy: further observations on blood loss from the gastrointestinal tract. *J. Pediatr.*, **116**, 11–18.

14 Advice for the middle-aged

J.S. Garrow

14.1 INTRODUCTION

What advice should be given about nutrition in the 1990s to the middle-aged? For the purpose of this meeting, 'middle age' means everyone between the children, about whom Professor Wharton spoke, and the elderly, about whom Dr Rosenburg will be talking. In particular, is there any reason why the advice which was given in the 1980s is not just as relevant for the next decade?

14.2 DEMOGRAPHY AND MORTALITY IN THE COMING DECADE

Two things are known about the population in the coming decade. People will be older than those in this country at present, and they will also be fatter. Of course, within the middle-aged age group, the age limits are fixed by definition, but during the 1980s the proportion of people who were in the 45–60-year age group decreased, while those in the 16 to 44 age group increased. During the next decade, demographic trends will ensure that these younger adults become older adults, so we will have more people to deal with in the 45 to 60 age range. From the viewpoint of public health, this is very important. The main causes of death among people aged 16 to 44 years are cancer, cardiovascular disease and respiratory disease. Among women in England and Wales aged 16–44 in 1987, cancer caused about 800 deaths per million, cardiovascular disease about 200 and respiratory disease about 100. Among men aged 16–44 in 1987 the mortality from cancer, cardiovascular disease and respiratory disease was about 600, 700 and 100 respectively.

When the age group 45 to 64 is considered, the same three diseases are the main cause of mortality, but the deaths per million are very different. Among women, the deaths from cancer are about 6000, from cardiovascular disease about 3000 and from respiratory disease about 500. Among men, these mortality rates are about 7000, 10 000 and 1000. It is obvious therefore that as people move into the 45 to 64 age group, the mortality from all these causes

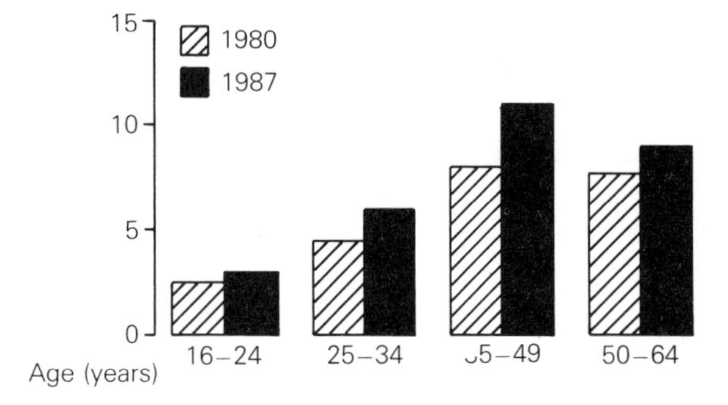

Figure 14.1 Percentage prevalence of obesity (Quetelet's index > 30) by age among a representative sample of British men in 1980 and 1987 (data of Rosenbaum *et al.*, 1985 and Gregory *et al.*, 1990).

increases and in particular there is a preponderance of mortality from cardiovascular disease among men.

14.3 THE INCREASING PREVALENCE OF OBESITY

The other change which must be anticipated in the coming decade is an increase in the prevalence of obesity. Figures 14.1 and 14.2 show the prevalence of obesity (Quetelet's index (QI) greater than 30) among men and women in the UK. Data are based on surveys of representative samples of the population between the ages of 16 and 64, which were done in 1980 (Rosenbaum *et al.*, 1985) and again in 1987 (Gregory *et al.*, 1990). It is

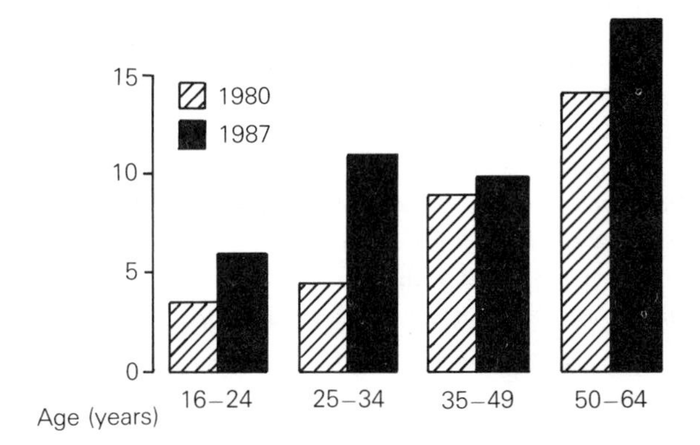

Figure 14.2 Percentage prevalence of obesity (Quetelet's index > 30) by age among a representative sample of British women in 1980 and 1987 (data of Rosenbaum *et al.*, 1985 and Gregory *et al.*, 1990).

evident that in both sexes and in all age groups there is an increase in the prevalence of obesity over this interval of 7 years. Both the prevalence and the increase in prevalence is greater among women. It is particularly worrying that among women aged 25 to 34 years, the prevalence of obesity is now over 10%, whereas in 1980 it was less than 5%.

The significance of this increase in obesity on public health is most clearly seen when we study the effect of obesity on the prevalence of diabetes. Bonham and Brock (1985) analysed the prevalence of diabetes in a very large sample of American people. The prevalence increased with age. It was higher among females than among males, and it was higher in black than in white people. However, within each age, sex and racial group there was a highly significant increase in the prevalence of diabetes with increasing obesity. Diabetes is not a very important cause of death, but a similar association between obesity and the prevalence of cardiovascular disease can be found. Hubert *et al.* (1983) has shown that in the Framingham population the incidence of myocardial infarction, congestive heart failure and all types of cardiovascular disease is highly significantly associated with obesity. This association remains significant even when other important predictors of cardiovascular disease such as age, cigarette smoking, blood pressure and serum cholesterol have been taken into account. Indeed, among women obesity is third in ranking as a predictor of cardiovascular disease, with only age and blood pressure being more strongly predictive.

The other important cause of death which should be considered is cancer. Some types of cancer, such as of the lung, are mainly related to cigarette smoking. However, certain sex-hormone-sensitive cancers are statistically more prevalent among obese people. In men, these are cancers of the colon, rectum and prostate and in women cancers of the breast, ovary, endometrium and cervix (Garfinkel, 1985). It is now known that the increased activity of aromatase in adipose tissue of obese people causes an increased conversion of androgens to oestrogens. This conversion probably explains the increased susceptibility to these cancers. Adipose tissue is also a large reservoir of cholesterol. Obese people are therefore more liable to form gall stones, because their bile is super-saturated with cholesterol.

All these penalties of obesity increase with weight gain and decrease with weight loss, with the exception of the risk of gall stone formation during weight loss in obese people (Liddle *et al.*, 1990). The cholesterol in adipose tissue is mobilized and the bile may become even more super-saturated.

14.4 DO THE NACNE (1983) GUIDELINES STILL APPLY?

To what extent do these differences between the population in the 1990s and the 1980s affect the nutritional advice which should be given? The author believes that most of the advice which was given in the NACNE (1983)

Table 14.1 NACNE (1983) revised for the 1990s

	NACNE (1983)		*Revision for the 1990s?*
	For the 1980s	*Long term*	
Obesity	Balance intake with more exercise		QI > 25 eat less, more exercise for all
Total fat as % E	34	30	30–33%
Saturated F A	15	10	15 decreasing to 12%
Polyunsat Wated F A	5		8%
Monounsat Wated F A			10–20%
Sucrose	20 kg/y (12% E)		Not >15% E
Fibre	25 g/d	30 g/d	25 g/d (6 g soluble)
Salt	3 g (=50 mmol/d)		<120 mmol/d
Alcohol	5% E	4% E	4% E

Abbreviations: E = energy, FA = fatty acids, QI = Quetelet's index

document still applies, but perhaps the emphases should be changed in the way suggested in Table 14.1. It is clearly wise to restrict the intake of fat, and in particular saturated fat, but it is suspected that few nutrition educators expect to reduce total fat intake to 30% of non-alcohol energy intake and would be prepared to settle for 33%, which is a considerable decrease in relation to present dietary practice. However, it seems that the public is willing to change to unsaturated fat, so a target of 8% polyunsaturated fat may be achievable. The contribution of mono-unsaturated fatty acids in reducing cholesterol is better accepted now than it was in 1983.

The long-term NACNE targets of 12% energy from sugar and an intake of 30 g dietary fibre may also be pushing the limits of acceptability further than the probable benefit would justify: it is considered by some workers that not more than 15% energy from sugar and a dietary fibre intake of 25 g, of which about a quarter should be soluble fibre, represents an attainable target for the 1990s which would be justified by the health benefit it would bring.

Undoubtedly far more sodium is eaten than required, and consumers rely mainly on food manufacturers to produce lower sodium formulations. It is doubtful if the NACNE target of 3 g/day (50 mmol/day) is justified or attainable, but a target of less than 120 mmol/day might be achieved with help from the food industry. Salt intake is of importance not only in relation to hypertension but also in view of the detrimental effects on calcium balance of a high sodium intake. At this meeting, evidence has been presented that sodium is not necessarily the main dietary electrolyte concerned with hypertension. A recent publication from India indicated that a daily supplement of 60 mmol potassium lowered blood pressure in mildly hypertensive patients but that magnesium (20 mmol/day) had no additive effect (Patki *et al.*, 1990). The advice to decrease average population energy intake from alcohol to 4% is justified, but of course this should be done by persuading those individuals with high alcohol intakes to stay within the limits prescribed by health educators.

14.5 RECENT CONCERNS NOT COVERED BY NACNE PROPOSALS

In general, NACNE was prepared to accept the 1979 recommended dietary allowances (RDAs) as an adequate guide to vitamin and mineral intakes. These RDAs are currently under revision and it would be unwise to make quantitative recommendations before that expert committee has reported. However, areas of concern can be noted. It is obvious that an RDA for iron is rather meaningless, since the amount of iron available to the individual depends, among other things, on the proportion of readily absorbed haem iron and on the intakes of vitamin C and protein which accompany the intake of iron-containing plant foods.

Antioxidants, and their role in scavenging free radicals, are now an important topic for nutritional debate. It cannot be assumed that the optimal intake of antioxidant vitamins is that amount which prevents the classical deficiency states, but there are still no good data about the advantages or disadvantages of very high vitamin intakes. In the last decade psuedo-nutrition has become big business, and the public is being offered a range of dietary supplements which have not been shown to do good, and which may very well do harm.

It is now evident that an important part of the prophylaxis of senile osteoporosis is to achieve a good skeletal mass in middle age, which in turn has implications about the calcium balance of children and young adults. Apart from an adequate calcium intake this involves the appropriate intake of vitamin D (or exposure to sunlight), the limitation of nutrients which decrease absorption of calcium (like phytate) or increase its excretion (like sodium), exercise, hormonal status and cigarette smoking. No doubt these are points on which the new RDA committee will express an opinion.

In an update of nutrition guidelines for Europe, James (1988) lists a fluoride concentration in water of 0.7–1.2 ppm as a nutrient goal. Despite the protests of anti-fluoriders, the case for water fluoride decreasing dental caries is overwhelming, and it is doubtful if it is possible to provide adequate protection to all the population by advocating the use of fluoride toothpaste, which is much more likely to cause overdose than water fluoridation.

14.6 PRIORITY FOR THE PROBLEM OF OBESITY

The aspect of advice which the author considers understressed in the NACNE document concerns obesity. The advice to balance energy intake and output is really not enough. It is important that people are aware of the range of weight for height which is associated with longevity and good health. Health professionals should be prepared to assist people who are above the optimal range to reduce their weight in a nutritionally sound manner. Obviously the prime targets for reduction are items of food such as sugar and alcohol, which provide energy but no other nutrients. This is not the place to describe the available methods for treating or preventing obesity, which have been reviewed in some detail elsewhere (Garrow, 1988), but certainly in the 1990s more attention should be given to the problem of preventing obesity.

It would be easier to prevent obesity if it were possible to identify a particularly high-risk group at whom preventive measures could be targeted, but everyone is susceptible to some extent. The only people who have a markedly increased risk of becoming obese are those who were formerly obese and who have lost a substantial amount of weight. This should not be interpreted as evidence that there is anything special about the genetics or metabolism of obese people: no doubt ex-smokers are more liable to start smoking than life-long non-smokers. About one in three adults in our country is already overweight (QI > 25), so a strategy for preventing obesity needs three components: first, community-based slimming clubs where those who wish to lose weight can obtain competent advice and support; second, a revision of public opinion about obesity which would assist those who have lost weight to maintain their weight loss; and third, a system which would enable those children entering primary school who are already overweight-for-height to receive a modified diet and exercise patterns to ensure that they grow more slowly in weight than in height over the next 5 years or so. They will thus enter their 'teens' with normal body proportions. Again, this is a matter which cannot be discussed in detail here, but it requires the intelligent co-operaion of parents, teachers, school-meals organizers, school doctors and community dietitians.

The data which have been presented indicate that obesity is an important problem in our country and one which is getting rapidly worse. The prime aim for the 1990s is that the trend towards increasing obesity in our population should be reversed.

REFERENCES

Bonham, G.S. and Brock, D.B. (1985) The relationship of diabetes with race, sex, and obesity. *Am. J. Clin. Nutr.*, **41**, 776–83.

Garfinkel, L. (1985) Overweight and cancer. *Ann. Inter. Med.*, **103**, 1034–6.

Garrow, J.S. (1988) *Obesity and Related Diseases*, Churchill Livingstone, London, p. 329.

Gregory, J., Foster, K., Tyler, H. and Wiseman, M. (1990) *The Dietary and Nutritional Survey of British Adults*, HM Stationery Office, London.

Hubert, H.B., Feinleib, M., McNamara, P.M. and Castelli, W.P. (1983) Obesity as an independent risk factor for cardiovascular disease: a 26-year follow-up of participants in the Framingham heart study. *Circulation*, **67**, 968–77.

James, W.P.T. (1988) *Healthy Nutrition: Preventing Nutrition-related Diseases in Europe*, European series No. 24, WHO, Copenhagen.

Liddle, R.A., Goldstein, R.B. and Saxton, J. (1990) Gallstone formation during weight-reduction dieting. *Arch. Intern. Med.*, **149**, 1750–3.

NACNE (1983) *Proposals for Nutritional Guidelines for Health Education in Britain*, Health Education Council, London.

Patki, P.S., Singh, J., Gokhale, S.V., *et al.* (1990) Efficacy of potassium and magnesium in essential hypertension: a double-blind, placebo controlled, crossover study. *Br. Med. J.*, **301**, 521–3.

Rosenbaum, S., Skinner, R.K., Knight, I.B. and Garrow, J.S. (1985) A survey of heights and weights of adults in Great Britain. *Ann. Hum. Biol.*, **12**, 115–27.

15 Nutritional requirements of the elderly

I.H. Rosenberg

15.1 INTRODUCTION

Nutritional issues should be addressed not only for the 1990s but also for the 21st century. A remarkable demographic change which has occurred over this century will accelerate into the next. In America now, as in many European countries, those over the age of 65 are about 1 in 9 of the population. Early next century they will be 1 in 5, a number which has already been achieved in Japan. It is generally agreed that this change in one century, from 1 in 25 of the population over age 65 in 1900 to 1 in 5 early in the 21st century, represents perhaps one of the most remarkable demographic changes that have been experienced in modern history and presents an enormous health challenge. Many of the conditions that afflict the elderly have been discussed in this meeting. Almost any kind of physiological function, whether it be renal concentrating function, pulmonary function, cardiac output or glucose tolerance, has been shown in cross-sectional studies to decline over the age span. Thus one clear goal of medical science should be to prevent, to the extent possible, those diseases and/or functional declines which contribute to debility and dependence in the elderly. Thus, nutritionists are therefore reflecting on the ageing process itself and the question of how many of these declining functions are, in fact, inevitable. The biological lifespan is probably fairly well defined by species and genetic heritage. Nutritionists ought therefore to be considering whether we can achieve a lifestyle of habit and diet so that physiological functions are retained through the span of age. The primary aim should therefore be to strive for an elderly population that is more vigorous, more active and more independent.

Few studies have directly addressed dietary goals and dietary requirements for the elderly. Researchers have been dependent on extrapolations of recommendations for younger adults. This reflects the fact that so many studies, including epidemiological studies about the relationship of diet and heart disease, have been carried out in young and middle-aged populations. The information available on the elderly, especially that concerned with dietary requirements, is still somewhat rudimentary. In this setting, comments on some of the kinds of current studies at the USDA Human Nutrition Center on Aging at Tufts University where metabolic studies are largely directed specifically to studies in the elderly are relevant. Researchers are now in a position to begin to address some of the questions about what is different about older subjects and what can be used as a basis for working towards specific dietary recommendations for an older population.

15.2 BODY COMPOSITION

First there is the issue of body composition. If data from cross-sectional studies are used, there is a striking decline in lean body mass which occurs over the decades of adult life (Cohn *et al.*, 1980). In females it is apparent that the most dramatic change appears after the menopause, as with declining bone mass (Riggs *et al.*, 1981). In the health sciences, decline in bone mass has received much more emphasis than the change in lean muscle mass. This decline in lean body mass is associated with a striking change in body composition as the ratio of lean muscle mass to fat declines. There is an associated decline in energy requirement over the age span which has two components. One is a decreasing basal energy requirement, which is very closely linked to the decline in lean body mass. There is an even more dramatic decline in energy intake needed to support diminishing physical activity. If the problem of obesity in the middle-aged and older population is to be addressed, not only the dietary intake but the physical activity, which is the other side of the energy balance equation must be considered. Of course, the increased sedentary behaviour of this population has a great deal to do with the changes in body composition and changes in obesity rates which have been experienced.

15.3 PHYSICAL ACTIVITY

Studies in the Human Nutrition Research Center's physiology laboratory show that, at age 65, a vigorous athlete can have the same body composition in terms of lean mass versus fat mass as the normal 35-year-old. It has now been found that even the frail, 90-year-old women who are in nursing homes can, with directed exercise, increase their muscle function by as much as 200% or 300% and even increase their muscle mass (Fiatarone *et al.*, 1990). Therefore, it is questionned whether the reported decline in lean body mass with ageing is, in fact, an inevitable result of ageing or whether it has to do in part with the great tendency in Western nations to slow down physical activity with age. The failure to maintain a lean mass probably reflects this physical inactivity more than the passage of time.

If nutritionists recommend continued physical activity to maintain lean body mass, will this have an impact on other nutrient requirements? This centre has performed some recent studies in respect to protein requirements in actively exercising subjects. Not until actively exercising subjects ingested more than current recommended levels for protein was positive nitrogen balance achieved (Meredith *et al.*, 1987). An interesting dilemma is whether a slightly higher protein RDA should be recommended for the elderly who are physically active in order to maintain this lean body mass. That will, of course, direct studies into some other issues about a dietary protein and declining renal function.

15.4 MAGNESIUM

Magnesium has been mentioned earlier. Studies by colleagues in some elderly in nursing homes indicate that one of the dietary components that have been found to correlate with muscle strength is magnesium (Fiatarone, personal communication). As suggested earlier by Dr McCarron, magnesium may be a problem nutrient in the advanced elderly. This deserves a good deal more study as an issue in the elderly, not just for the question of its relationship to blood pressure but also in terms of muscle mass and strength.

15.5 VITAMIN D

Another nutrient which deserves concern in respect to muscle strength as well as bone mass in the elderly is vitamin D. Levels of 25-hydroxy vitamin D could be plotted which also correlate with muscle strength in elderly individuals, with the least strong having the lowest 25-hydroxy vitamin D levels (Fiatarone, personal communication). It is known that there is a decline in calcium absorption over the age span (Avioli *et al.*, 1965). Participants heard earlier about the possibility that this might be related to a reciprocal increase in blood pressure. Causes of that decline are multiple. 1,25-dihydroxy vitamin D levels decline over the age span as do 25-dihydroxy vitamin D levels (Orwoll and Meier, 1986). It was found that in a nursing home population the levels were commonly in a relatively low range (Fiatarone, personal communication).

Vitamin D metabolite levels are low for several reasons. First, this elderly population has very low vitamin D intakes. Second, they have a diminished capacity to form vitamin D in the skin, and third, they synthesize 1,25-

dihydroxy vitamin D in the kidney from precursors at a reduced rate. In the Netherlands, at a latitude not dissimilar from Scotland's, there is a strong relationship between the seasonal hours of sunlight and the levels of serum 25-hydroxy vitamin D in adults (Lips *et al.*, 1982). In Boston, which is lower in latitude than Edinburgh, it has been found that for at least three or four of the winter months there is no vitamin D formed in the skin of subjects (Webb *et al.*, 1988) and that therefore the only time that sunshine is really effective in producing vitamin D is in the summer months. The author is confident that the same phenomenon occurs in the UK. The elderly have an additional problem. It should be remembered that vitamin D in the skin is formed from 7-dehydrocholesterol. The efficiency of this process decreases with age. The elderly have less 7-dehydrocholesterol in the skin and it is converted, in the presence of the same amount of sunlight, less efficiently (Holick, personal communication). Finally, the aged intestine is less responsive to the calciotopic effects of 1,25-dihydroxy vitamin D (Armbrecht, 1988).

There are, therefore, a number of reasons why the elderly are more likely to be vitamin D-deficient and less likely to utilize vitamin D effectively. What should we be doing to meet these additional requirements in view of bone disease and muscle function and other trophic effects of vitamin D? This may be one area where supplementation needs to be considered when dietary sources of vitamin D are limited.

15.6 PYRIDOXINE

Vitamin B_6 is another vitamin which is believed to be a problem nutrient in the elderly. If plasma pyridoxal phosphate is measured at base line and older individuals are placed on a low B_6 diet, as was recently done at the HNRC, the drop in the circulating B_6 levels on the low intake can be observed and the response to increasing levels of repletion can be measured. When colleagues repleted these elderly subjects with a full RDA dose as established for young adults, they were rarely able to restore fully the circulating pyridoxal phosphate levels (Ribaya-Mercado *et al.*, 1990). They found that B_6 doses above the RDA were needed in order to restore fully not only the circulating vitamin levels but the subtle decline in brain function which was documented electrophysiologically, although not behaviourally.

15.7 IMMUNE FUNCTION

Another observation that came out of this study was related to immune function (Meydani *et al.*, 1990). Whether lymphocyte proliferation or interleukin production was assessed, at the nadir of the B_6-deficient diet there was a progressive increase in these parameters over the course of repletion such that these measurements could have been used as surrogates for B_6 status.

15.8 ANTIOXIDANT VITAMINS

It is known that immune function declines over the age span. How much of that decline in immune function is in fact related to adequacy or inadequacy of diet or micronutrients continues to be a very interesting challenge. Dr Simin Meydani of the Nutritional Immunology Laboratory performed skin tests on volunteers in the 65 to 70 year category. She was studying at the same time the issue of antioxidant nutrients as related to immune function. When older subjects were put either on placebo or vitamin E supplement, a remarkable stimulation occurred in the reactivity of the skin test in the supplemented group (Meydani *et al.*, 1989). The effects of an antioxidant nutrient in high doses, many times the RDA, have been observed. The implications of this observation for a dietary antioxidant which might have stimulatory effects on immune function, one of those functions which declines over the life span, is not yet known.

15.9 DIGESTIVE FUNCTION

As a gastroenterologist, I would like to consider digestive function in the elderly as a factor in nutrient requirements and metabolism. One of the striking changes which occurs in the gastrointestinal tract with ageing is the loss of the ability to produce stomach HCl as an expression of the appearance of atrophic gastritis. Those with atrophic gastritis and achlorhydria can now be identified with a very high degree of accuracy by measuring the pepsinogen I and II levels in blood (Russell *et al.*, 1986). The nutrition status survey of Boston elderly found that, by the age of 70, a quarter of the people had lost their ability to make stomach acid, a third by the age of 80, and 40% beyond 85 (Russell *et al.*, 1986). I am not sure whether prevalence studies are available in this country, but I suspect they would not be much different from those of the United States. Prevalence of atrophic gastritis is actually rather higher in Finland (Samloff, personal communication). It is suggested that this finding has significant effects on some physiological and absorptive functions. For example, the vitamin B_{12} levels in young adults are higher than in normal elderly and lowest in adults with atrophic gastritis (Russell *et al.*, 1987). This is probably because in the absence of acid, there is a problem with the absorption of food B_{12}. If moderate atrophic gastritis patients are compared with controls, the former can absorb crystalline vitamin B_{12} normally but when one administers vitamin B_{12} associated with protein as in the diets, those with atrophic gastritis are much less able to absorb (Russell *et al.*, 1987). The same phenomenon exists in the case of folic acid. Patients who have atrophic gastritis have low folic acid absorption rates (Russell *et al.*, 1979). When HCl was added to replace stomach acid, folate absorption was normal. Thus there are two vitamin deficiencies in which the elderly are potentially at risk with increasing atrophic gastritis.

15.10 HOMOCYSTEINE

Based on recent technical developments, total plasma homocysteine can now be used as an additional marker of some of these problems (Kang *et al.*, 1979). It will be remembered that homocysteine is converted to methionine in the presence of vitamin B_{12} and methyl folate. Metabolism of homocysteine to cystathionine requires vitamin B_6. Therefore, any one of these three vitamin deficiencies will result in a homocysteine elevation in blood. Elevated homocysteines will fall with treatment with vitamin B_{12} (Lindenbaum *et al.*, 1988), folate (Brattstrom *et al.*, 1988) and/or vitamin B_6 (Brattstrom *et al.*, 1990). About 15% of elderly subjects at the HNRC have elevated levels of homocysteine. There is concern about homocysteine beyond the fact that it may be a marker of vitamin deficiency, because elevated homocysteine levels are associated with cardiovascular disease (Uelund and Refsum, 1989). In homocystinuria, a genetic defect, premature vascular disease is prominent. Elevated homocysteine appears to predispose to toxic effects in the vascular system (McCully, 1969). HNRC investigators compared the distribution of homocysteine levels in a control population with the distribution in a population with premature coronary artery disease (Genest *et al.*, 1990). With coronary artery disease, the homocysteine distribution was shifted to higher levels. Some investigators have suggested that homocysteine may explain some of the risk of cardiovascular disease that is unexplainable by lipoprotein and cholesterol parameters.

15.11 CATARACT

Cataract is a very common problem in the elderly, worldwide. We have found in our population studies that there is a three-fold higher risk of central cataract in those with low vitamin C intake or status compared with those with high vitamin C status and an even more striking 11-fold difference in respect to peripheral cataract (Jacques *et al.*, 1988). There are more modest differences if plasma carotenoids or vitamin E are investigated (Jacques *et al.*, 1988). These observations add support to studies showing that antioxidant nutrients may protect against cataract formation (Taylor, 1989).

15.12 SUMMARY

Some problem nutrients in the elderly have been identified. Detailed recommendations to meet specific dietary requirements have not been made. It is considered that all of these nutrients may need to be addressed in respect to dietary needs of the elderly. Among mineral elements, calcium and magnesium are mentioned, but there is some evidence that zinc and chromium are also potential problem nutrients in the elderly. One of the real challenges in the 21st century will be to set standards to meet some of these special requirements of our ageing populations.

REFERENCES

Armbrecht, H.J. (1988) Changes in the components of the intestinal calcium transport system with age, in *Aging in Liver and Gastrointestinal Tract* (eds L. Bianchi, P. Holt, O.F.W. James and R.N. Butler), MTP Press Boston.

Avioli, L.V., McDonald, J.E. and Lee, S.W. (1965) The influence of age on the intestinal absorption of ^{47}Ca in post-menopausal osteoporosis. *J. Clin. Invest.*, **44**, 1960–7.

Brattstrom, L.E., Israelsson, B., Jeppsson, J.-O. and Hulterg, B.L. (1988) Folic acid–an innocuous means to reduce plasma homocysteine. *Scand. J. Clin. Lab. Invest.*, **48**, 215–21.

Brattstrom, L., Israelsson, B., Norrving, B., *et al.* (1990) Impaired homocysteine metabolism in early onset cerebral and peripheral occlusive arterial disease. *Atherosclerosis*, **81**, 51–60.

Cohn, S.H., Vartsky, D., Yasumura, S. *et al.* (1980) Compartmental body composition based on total body nitrogen, potassium, and calcium. *Am. J. Physiol.*, **239**, E524–E530.

Fiatarone, M.A., Marks, E.C., Ryan, D.N., *et al.* (1990) High-intensity strength training in nonagenarians. *JAMA*, **263**, 3029–34.

Genest, J.J., McNamara, J.R., Salem, D.N., *et al.* (1990) Plasmahomocysteine levels in men with premature coronary artery disease. *J. Am. Coll. Cardiol.*, **16**(5), 1114–9.

Jacques, P.E., Hartz, S.C., Chylack, L.T., *et al.* (1988) Nutritional status in persons with and without senile cataract: blood vitamin and mineral levels. *Am. J. Clin. Nutr.*, **48**, 152–8.

Kang, S.-S., Wong, P.W.K. and Becker, N. (1979) Protein-bound homocysteine in normal subjects and in patients with homocystinuria. *Pediat. Res.*, **13**, 1141–3.

Lindenbaum, J., Healton, E.B., Savage, D.G., *et al.* (1988) Neuropsychiatric disorders caused by cobalamin deficiency in the absence of anemia ormacrocytosis. *N. Engl. J. Med.*, **318**, 1720–8.

Lips, P., Hackeng, W.H.L., Jongen, M.J.M., *et al.* (1982) Seasonal variation in serum concentrations of parathyroid hormone in elderly people. *J. Clin. Endocrinol. Metab.*, **57**, 204–6.

McCully, K.S. (1969) Vascular pathology of homocysteinemia: implications for the pathogenesis of arteriosclerosis. *Am. J. Pathol.*, **56**, 111–28.

Meredith, C.N., Zackin, M.J., Frontera, W.R. and Evans, W.J. (1987) Body composition and aerobic capacity in young and middle-aged endurance-trained men. *Med. Sci. Sports Exerc.*, **19**, 557–63.

Meydani, S.N., Meydani, M., Barklund, P.M., *et al.* (1989) Effect of vitamin E supplementation on immune responsiveness of the aged. *Ann. NY Acad. Sci.*, **570**, 283–90.

Meydani, S.N., Ribaya-Mercado, J.D., Russell R.M., *et al.* (1990) The effect of vitamin B_6 on the immune response of healthy elderly. *Ann. NY Acad. Sci.*, **587**, 303–6.

Orwoll, E.S. and Meier, D.E. (1986) Alterations in calcium, vitamin D, and parathyroid hormone physiology in normal men with aging: Relationship to the development of senile osteopenia. *J. Clin. Endocrinol. Metab.*, **63**, 1262–9.

Ribaya-Mercado, J., Russell, R.M., Sahyoun N., *et al.* (1990) Vitamin B_6 deficiency elevates serum insulin in elderly subjects. *Ann. NY Acad. Sci.*, **585**, 531–3.

Riggs, B.L., Wahner, H.W., Dunn, R.B., *et al.* (1981) Differential changes in bone mineral density of the appendicular and axial skeleton with aging. *J. Clin. Invest.*, **67**, 328–35.

Russell, R.M., Dhgar, G.J., Dutta, S.K. and Rosenberg, I.H. (1979) Influence of intraluminal pH on folate absorption: studies in control subjects and in patients with pancreatic insufficiency. *J. Lab. Clin. Med.*, **93**, 428–36.

Russell, R.M., Krasinski, S.D., Samloff, I.M., *et al.* (1986) Folic acid malabsorption in atrophic gastritis: compensation by bacterial folate synthesis. *Gastroenterology*, **91**, 1476–82.

Russell, R.M., Suter, P.M. and Golner, B. (1987) Decreased bioavailability of protein bound vitamin B_{12} in mild atrophic gastritis: reversal by antibiotics. *Gastroenterology*, **92**, 1606.

Selhub, J., Dhar, G.J. and Rosenberg, I.H. (1983) Gastrointestinal absorption of folates and antifolates. *Pharmacol. Ther.*, **20**, 397–418.

Taylor, A. (1989) Associations between nutrition and cataract. *Nutr. Rev.*, **47**, 225–34.

Uelund, P.M. and Refsum, H. (1989) Plasma homocysteine, a risk factor for vascular disease: plasma levels in health, disease, and drug therapy. *J. Lab. Clin. Med.*, **114**, 473–501.

Webb, A.R., Kline, L. and Holick, M.F. (1988) Influence of season and latitude on the cutaneous synthesis of vitamin D_3: exposure to winter sunlight in Boston and Edmonton will not promote vitamin D_3 synthesis in human skin. *J. Clin. Endocrinol. Metab.*, **61**, 373–8.

16 Nutrition in the nineties: an overall view

J. Hautvast

16.1 INTRODUCTION

The objective of this conference is to evaluate nutrition in Europe in the nineties. When discussing this issue nutritionists should be clear about the goals for nutrition in the future. Do all of us as nutritional scientists in Europe think the same about such goals?

- Are there a number of general goals or should very specific goals be formulated?
- Which activities or programmes should be implemented to reach defined goals?
- Which factors determine the success or failure of the realization of nutritional goals?
- Which assumptions could be made within this context?
- Does a good system to monitor and evaluate progress in reaching nutritional goals exist or should one be developed?

It is not possible to discuss all of these questions in great detail. This contribution will reflect on these issues and the points raised which it is hoped will be of relevance in a discussion on nutrition in the nineties.

16.2 NUTRITIONAL GOALS FOR EUROPE IN THE NINETIES

When reviewing nutritional guidelines and nutritional goals in different European countries one may easily come to the conclusion that the following dietary guidelines may be the most important ones in the coming decade:

- eat a wide variety of foods;
- maintain an optimal body weight;
- reduce fat intake, especially of saturated fat;
- increase carbohydrate intake;

- decrease sugar, salt and cholesterol intakes;
- increase dietary fibre intake;
- avoid excessive alcohol intake.

Such dietary guidelines are very much in line with the interim dietary guidelines formulated by the European Community in 1987 when launching the programme 'Europe against Cancer'. The action plan of the programme is composed of what are termed ten simple rules, four of which are concerned with nutrition:

- consume alcoholic drinks in moderation;
- eat sufficient fresh fruits and vegetables;
- eat sufficient cereals with a high dietary fibre content;
- eat low fat foods and avoid becoming overweight.

In a publication of the WHO Regional Office for Europe in Copenhagen (James *et al.*, 1988), intermediate and ultimate nutrient goals for Europe were formulated based upon information on the causes and the prevalence of a number of nutrition-related diseases which could be regarded as preventable by adequate dietary measures. These goals are presented in Table 16.1.

In addition to the goals presented in Table 16.1, the following are both

Table 16.1 Intermediate and ultimate goals for the intake of nutrients in Europe (from James *et al.*, 1988)

	Intermediate goals		*Ultimate goals*
	General population	*Cardiovascular high-risk group*	
Proportion (%) of total energy[a] derived from:			
complex carbohydrates[b]	>40	>45	45–55
protein	12–13	12–13	12–13
sugar	10	10	10
total fat	35	30	20–30
saturated fat	15	10	10
P:S ratio[c]	<0.5	<1.0	<1.0
Dietary fibre[d] (g/day)	30	>30	>30
Salt (g/day)	7–8	5	5
Cholesterol (mg/4.18 MJ)	—	<100	<100
Water fluoride (mg/l)	0.7–1.2	0.7–1.2	0.7–1.2

[a] All values given refer to alcohol-free total energy intakes
[b] Figures for complex carbohydrate are derived from the other recommendations
[c] Ratio of polyunsaturated to saturated fatty acids
[d] Dietary fibre values are based on analytical methods that measure non-starch polysaccharides and enzyme-resistant starch produced by food processing or cooking

intermediate and ultimate nutrient goals for the general population and the high-risk group:

- limitation of alcohol intake;
- application of iodine prophylaxis when necessary;
- increase in nutrient density;
- a body mass index (BMI) of 20–25, although this value is not necessarily appropriate for developing countries where the average BMI is probably 18.

It should be realized, however, that generally such guidelines are not considered to be very exciting, new or unique since they have been discussed for years or even decades. People are tired of this subject and there is hardly any lasting enthusiasm to work hard and consistently in order to change dietary habits. The reason for such an attitude is not clear and should be analysed. Questions such as the following can then be raised:

- is the scientific evidence reliable and sufficient to back the dietary changes;
- are the appropriate professional people sufficiently involved in programmes to change dietary habits;
- what roles do organizations such as the food industry play in this issue;
- and, how relevant is a change in agricultural policies?

It is obvious that present and future nutritional research findings may cause both excitement and doubts about dietary guidelines. However, subjects such as nutritional policy and nutritional education lead neither to excitement nor to doubt, but more often to frustration. Therefore the issues involved will be examined in more detail.

16.3 NUTRITIONAL RESEARCH IN EUROPE

Most nutritionists would agree that Europe needs strong nutritional research groups in order to identify and analyse continuously present and future questions concerning nutrition-related diseases. Findings of such research groups determine policies and programmes in nutrition in Europe. The author has recommended previously (Hautvast, 1990) that it is important to review and evaluate the extent and quality of nutritional research in Europe. Although there are increasing opportunities for nutritional research in Europe, continuation of such research is only guaranteed when it is of a high standard. An evaluation of on-going research may lead to recommendations that some nutritional topics require more attention in the future.

Nutritional research in recent years, especially in Europe, has given us surprising new findings which are of relevance in specifying future nutritional goals. One example of such research is the issue of the relationship of body weight to health and disease. In the dietary guidelines, mention is made of maintaining optimal body weight with a BMI (weight/height^2) between 20 and 25 (James *et al.*, 1988). However, research in the last decade carried out by people such as Per Björntorp (Göteborg) and Jaap Seidell (Wageningen) indicate that measures of body fat distribution may be more specific than information on body weight in defining risk of disease. In what way will such findings be introduced into dietary guidelines?

Studies of the role of fat intake in cholesterol metabolism, which is important for understanding the relationship between nutrition and cardiovascular diseases, have brought us intriguing new findings in recent years. A leading role in Europe has been played by the Wageningen researchers Martijn Katan, Ronald Mensink, Anton Beynen and Clive West. Recently the first two investigators (Mensink, 1990) reported unique findings on the role of both *cis*- and *trans*-monounsaturated fatty acids. *Trans*-monounsaturated fatty acids are found in ruminant fats and are produced during the industrial hydrogenation of polyunsaturated fatty acids. Very little information was available about the role of these fatty acids and such information as was available was controversial.

The studies were designed as follows.

- *Cis*-monounsaturated fatty acids (mainly oleic acid) were compared with carbohydrates. According to Keys' equation, both nutrients should have a similar effect on serum total cholesterol levels, but specific effects on the level of cholesterol in high density lipoproteins (HDL) or low density lipoproteins (LDL) were unknown.
- *Cis*-monounsaturated fatty acids were compared with polyunsaturated fatty acids of the $(n-6)$ family. It was thought that polyunsaturated fatty acids lower serum cholesterol levels more effectively than monounsaturated fatty acids. It was unclear whether the difference, if any, was due to the lowering of the level of cholesterol in the HDL and LDL fractions of serum.
- The most recent study was on *trans*-monounsaturated fatty acids which are consumed in large amounts as hydrogenated oils and fats and with no information on lipoprotein levels.

The findings of these studies have been published in various international journals and were briefly summarized in the thesis of Mensink (see Table 16.2). From these findings it may be concluded that *cis*-monounsaturated fatty acids have the same effect on HDL cholesterol levels as polyunsaturated fatty acids and that *trans*-monounsaturated fatty acids produce an unfavourable distribution of cholesterol over the various lipoprotein fractions. Such new information is very important both as an achievement in nutritional research and from the point of view of influencing food choice (*cis*- or *trans*-monounsaturated versus polyunsaturated fatty acids). Such information is also important for food technologists in the context of hydrogenated fatty acids.

Table 16.2 Effects of substituting various fatty acids in the diet on the concentration of cholesterol in lipoprotein fractions and of triglycerides (from Mensink, 1990)

Substitution	*Cholesterol*				*Triglycerides*
	Total	*LDL*	*HDL*	*HDL/LDL*	
Cis-monounsaturated FA for carbohydrates	=	=	++	++	—
Cis-monounsaturated for polyunsaturated FA	=	=	=	=	=
Cis- for *trans*-monounsaturated FA	–	—	++	++	–
Cis-monounsaturated for saturated FA	—	—	=	++	–
Trans FA for saturated FA	–	–	—	–	=

FA = fatty acids

As far as nutritional research in Europe is concerned, two conclusions can be drawn. First, Europe needs continuous heavy investment in high quality research in order to obtain a better insight into the role of nutrients in health and disease. Second, future nutritional research findings may lead to more specific dietary guidelines. At the present time, such guidelines are still very general and as such difficult to apply.

16.4 FOOD AND NUTRITIONAL POLICY IN EUROPE

As indicated above, official food and nutritional policies, both at the national and European level, too often lead to feelings of frustration. Such policies laid down by official bodies are too general and often propose goals which are far too optimistic and too unrealistic. Of course, it is very exciting to be involved in activities regarding future scenarios and designing goals which it is hoped will lead to changes in dietary patterns and improvements in nutritional health. However, it must be said that achievements up until now have often been nil or very limited and have led to lack of interest and frustration by all concerned and more especially consumers.

A number of examples can be presented to illustrate this issue.

- In 1974, FAO organized the World Food Conference in Rome. At this congress it was officially stated by no less a person than Henry Kissinger on behalf of the President of the United States of America, that by the year 1985 no child would be seen with clinical signs of protein-energy malnutrition.

- In 1990 at the Regional FAO Conference in Venice, it was stated that Europeans are in need of a more balanced diet and the criteria and measures to achieve this goal appear to be very optimistic.
- In 1980, the WHO Regional Committee for Europe proposed a European regional strategy for attaining health for all by the year 2000. Target 16 is of specific relevance to people who are involved in nutrition, stating that 'By 1995, in all Member States, there should be significant increases in positive health behaviour, such as balanced nutrition, nonsmoking, appropriate physical activity and good stress management'.

Since 1980, two-thirds of the time period for the European 1995 goal of WHO have passed, and nothing special has occurred. It seems unlikely that in the remaining 5 years significant changes will take place in the direction of balanced nutrition.

In October 1990 the WHO Regional Office did organize the First European Food and Nutrition Policy Conference in Budapest. At the end of 1992 FAO and WHO, in collaboration with other international agencies, are planning to hold a World Nutrition Conference in Rome. This conference will be a follow-up of the World Food Conference, held in 1974 in Rome.

The statements formulated at the Budapest meeting are not different from those of other conferences. It is also to be expected that statements and resolutions formulated at the Rome meeting will not differ in principle from those of other conferences. Nevertheless, such conferences offer opportunities which nutritional scientists should grasp in order to provide an input into public policy. They give nutritional science and the professionals involved unique opportunities to bring their critical knowledge and rich experiences into the open. This should lead to the awareness that high quality nutritional science is very much needed in the future in order to formulate recommendations on nutritional, food and agricultural policies and on industrial processing practices required to produce foods which can enable consumers to meet nutritional guidelines.

16.5 RECOMMENDED DIETARY ALLOWANCES (RDAs) FOR EUROPE

An example of an urgent need for a common European approach is the formulation of European RDAs. In all EC countries, and probably in all European countries, RDAs have been established after often lengthy discussions by various national expert committees. It is idle, and indeed erroneous, to suggest that the national differences observed in the various tables of RDAs can be rationalized intelligently. There is absolutely no scientific reason to suggest that the nutrient requirements of Dutch or Greek people should differ significantly. By the time the single market is introduced at the end of 1992 a set of European RDAs should be available. It is important that nutrition professionals in Europe should provide leadership and take the prime responsibility for this subject. It is, however, not very clear along which lines this will take place.

16.6 ASSUMPTIONS TO ACHIEVE DIETARY GUIDELINES IN THE NINETIES

It has been stated above that on the one hand dietary goals are formulated in too optimistic a way, while on the other hand achievements in this field are often very disappointing. In order to achieve more, the issues which may play an important role should be analysed and discussed, for example under the headings listed below.

(a) Will new scientific evidence on nutrition-related diseases support dietary changes which have been proposed to promote good health?

It is difficult to predict whether this evidence will emerge and the extent of its influence. However, there is a growing awareness that general dietary guidelines are not very well received by consumers. More success could possibly be achieved if dietary guidelines were more tailored to the individual. Perhaps in several decades from now, each person will carry a DNA-nutrition passport indicating unfavourable hyper-responses to certain nutrients!

(b) Will changes in EC agricultural policies improve the nutritional quality of future diets?

Present agricultural policies, especially in EC countries, are under severe pressure caused mainly by economical and environmental problems. The growing awareness in the EC that agriculture policy should be based not solely upon quantity but also on quality might support improvements in the nutritional characteristics of future diets.

(c) What will be the role of the medical profession?

The medical profession has been neither much involved nor much interested in issues related to nutrition and health. However, changes seem to be occurring in this respect. In international medical journals such as the *New*

England Journal of Medicine and the *Lancet*, more scientific papers are appearing on health topics related to nutrition. Furthermore, it can be observed that special diets are more frequently used as a therapeutic approach alongside medicines. It may be stated that with an increasing involvement of the medical profession in the nutrition field, changes in dietary patterns may be achieved. The medical profession has probably an important influence on dietary habits and nutritional care.

(d) What will be the role of the food industry?

The food processing industry is showing a keen interest in broadening the range of food products offered to consumers. More and more products are manufactured with reduced energy, with less fat or with fats which are recommended from the health point of view. The food industry seems to be very keen to introduce so-called 'healthy' foods. Thus one may assume that the food industry will be supportive of changes in nutrition policy as it is recognized that consumers are increasingly looking for foods with a higher nutritive value.

(e) What attention will be given by the media to food and nutritional issues in the future?

At present, still too much emphasis is given to food faddism instead of scientifically-based food choices. Although scientific nutritional findings do achieve front-page coverage in both national and international newspapers and magazines, there will always be a dualism between balanced nutritional news and speculative food faddism news. The media are aware of consumers' interest in any information on foods. On the other hand, consumers seem to forget quickly what has been said about foods in the media.

(f) Will consumers be sufficiently aware and interested to discuss food issues and to follow diets which match the recommended dietary guidelines?

On the one hand it can be observed that consumers are showing a growing awareness of and interest in food issues. On the other hand, it should be realized that individual consumers do not consider themselves in direct need of food and nutrition messages. Consumers regularly report that their dietary pattern is adequate and that poor dietary habits are practised by other people and not by themselves. This issue obviously needs more study because any nutrition message should reach a broad group of consumers who are interested, willing and able to judge nutritional messages on their merits.

However, consumer behaviour needs careful study in the context of future scenarios and possible policies. Successes or failures in reaching dietary goals may very much depend on an adequate appraisal of the role of consumer behaviour.

16.7 NUTRITION PROFESSIONAL ORGANIZATIONAL STRUCTURE IN EUROPE

In the future more and more activities will be dealt with at a European level. This holds true for nutritional issues as, for example, the need for European RDAs. Within the context of Europe there is an obvious growing need for a professional organization of nutrition scientists. Such an organization could, for example, facilitate and stimulate cooperation and exchange in research, review and stimulate adequate nutrition training programmes, harmonize nutrition information and communication on a European scale and become an important counterpart for institutions such as the CEC and the WHO Regional Office.

The author has recently proposed that a European Institute of Nutrition (EIN) comparable to the American Institute of Nutrition (AIN) should be constituted and that links should be forged with AIN (Hautvast, 1990). The EIN should be an organization both complementary and closely linked to the Federation of European Nutrition Societies (FENS) and the Group of European Nutritionists (GEN). Membership of EIN should provide a focus for assurance of high quality.

16.8 FINAL REMARKS

Nutrition and health in Europe is a fascinating topic and has a bright future. In the past decade EC concerted actions such as EURONUT 'Nutrition and Health' have shown that there is a growing and keen interest in Europe to cooperate in nutrition, in both research and in training. The urge and necessity to cooperate is based on hypotheses and research findings regarding the role of nutrition in health and disease. As Europe has large variations in both dietary habits and morbidity and mortality patterns, unique possibilities are offered to study in more detail the important subject of the role of nutrition in health and disease. It is evident that the future challenge in Europe lies in nutrition training and research.

REFERENCES

James, W.P.T., Ferro-Luzzi, A., Isaksson, B. and Szostak, W.B. (1988) *Healthy Nutrition: Preventing Nutrition-related Diseases in Europe*, European Series No. 24, WHO Regional Publications, Copenhagen.

Hautvast, J.G.A.J. (1990) Nutrition and health in Europe: there is a future. *BNF Nutr. Bull.*, **15**, 12–20.

Mensink, R.P. (1990) Effect of monounsaturated fatty acids on high-density and low-density lipoprotein cholesterol levels and blood pressure in healthy men and women. PhD thesis, Wageningen Agricultural University, Wageningen, The Netherlands.

DISCUSSION

Parry, Edinburgh Professor Garrow, you have placed considerable emphasis on achieving a desirable relationship between weight and height. For many individuals this is not an easy matter, what practical advice could you give?

Garrow The first practical advice is to have reasonable information of a desirable range of weight and height. This is not always available. Commercial slimming clubs tend to quote slightly unrealistically low desirable weight, concealing the large range of weight over a 2 stones range. The other thing which militates against people attaining and maintaining a normal weight is the absurdly optimistic notion of possible rates of weight loss. Magic slimming cures suggest a loss of 10 lb per week and when that does not happen it leads to despondency. Sensible advice is a realistic target and a realistic rate of weight loss.

Lean, Glasgow You recommended reducing saturated fat, but you did not mention total fat. Is this an important issue that needs further discussion, should we be prioritizing saturated fat reduction rather than total fat? Is it appropriate to the middle-aged or should we be considering the development of heart disease from an earlier stage perhaps?

Garrow Yes, I was not being terribly novel in suggesting a reduction in saturated fat rather than total fat. On the available evidence, the advice to reduce in particular saturated fat seems to me to be a reasonable thing to do because if this actually comes about then of course the total fat intake will necessarily be reduced. The P/S ratio will necessarily increase because you are reducing saturated rather than polyunsaturated fat.

Davidson, Dundee Over the past few days I have listened with interest to all of the speakers, some of whom have mentioned the antioxidant theory. From the lack of discussion of the this topic, are we to assume that it is not controversial any more?

Garrow I think it is pre-controversial rather than post-controversial, if you see what I mean. Obviously advice about increasing intake of fruit and vegetables will have an impact on the antioxidants as well as other nutrients and soluble fibre. The same applies to some extent to the comments about reducing the intake of fat and maintaining body weight. As Brian Wharton rightly said 'If you are going to give advice you have really got to be pretty sure that you are doing good rather than harm'.

Williams, Guildford You mentioned, as have other speakers, the association between breast cancer and heavier body weight. In the pre-menopausal forms of breast cancer the increased risk is associated with leanness.

Garrow The cause of breast cancer when it is known is likely to be multifactorial. It was included on the list because it is plausible as a hormone-sensitive cancer with a reasonable and plausible connection between obesity and subsequent events via the aromatase activity in adipose tissue. The rate of androgen to oestrogen conversion has been shown to be very high in adipose tissue. The association with leanness I presume is a manifestation of the several paths in the aetiology of this malignancy. If you just take whole populations, which is what Garfinkle did, then it is more closely related to obesity than to leanness.

Campbell Brown, Glasgow Why has there been this massive increase in the proportion of women who are overweight in the middle age groups? Is it dietary or do you think it is other factors in people's lifestyle?

Garrow I really do not know, and of course what we do not know is whether what is being observed between 1980 and 1987 is simply a section of an existing trend as the 1980 survey was the first national representative survey. I cannot imagine that the metabolism of people has changed between 1980 and 1987 and therefore presumably the increased prevalence of obesity is due to an increase in total energy intake. It is unlikely to be something which started operating around 1980 and is more likely to be a continuous process that has been going for ages.

Winkler, London Could you expand on your justification for the recommendation to eat less sugar specifically in the middle age?

Garrow The justification for eating less sugar was that I regard obesity as being an important factor. I think sugar is a source of energy and nothing else and, therefore, the advice to reduce sugar will reduce energy intake without the loss of any other important nutrient. This, of course, differs from NACNE which said that there should be a reduction in sugar with a replacement by complex carbohydrate. Whether it is appropriate to replace it with complex carbohydrate depends on whether you are trying to maintain body weight or not. As I tried to point out to you, an increasing proportion of the population in this country should not be trying to maintain body weight, hence I omitted the usual rider about replacing it with complex carbohydrate. For those of you who are trying to maintain constant body weight, of course, replacement with complex carbohydrate is the appropriate thing to do.

Clarke, Glagow Professor Wharton do you have any views on the content of A, C and D vitamins, in the standard government vitamin drops which are presently issued. Do you think they need to be modified?

Wharton If you take the standard dose, which is now 5 drops daily, this gives you 7 microgrammes of vitamin D a day. This is certainly enough to prevent rickets, and even with additional amounts from other sources such as infant formula, it is unlikely to reach toxic levels. I always used to think that vitamin A was an unnecessary traveller. It was in the drops because the source of the vitamin D often included vitamin A as well. I suppose now we would have to be more cautious in view of the newer interests in vitamin A. However, the basic problem is do we need vitamin drops as supplements, and if so just when should they be used rather than the actual formulation?

Meehaf, Reading This is a question for Dr Rosenberg. I was not sure that the data presented really were an argument for increasing the RDAs, as it was not clear that those elderly subjects who were benefitting from the supplements had in fact been ingesting less than the current RDA to begin with. If they were not ingesting the current RDA then an increase would not be that helpful. Was their current condition caused by ingesting less than the current RDA?

Rosenberg The subjects were all ingesting at or above the current RDA, but over a range. That, in my judgment, is not a study which I think we can use in any quantitative way to address the issues of RDAs. I raised that point only to say that like some other kinds of physiological declines which are associated with ageing, the decline in the immune function is something that may not be an age-dependent phenonemon strictly, but may in fact be able to be modified by a nutritional or dietary or supplementary means. This is an example of the policy decision which we will have to face. In the short term, I think we should err on the side of a cautious recommendation for some of these antioxidant nutrients which may be important to immunity, to cataract prevention and so forth. But I think the dose, and whether this is achieved by RDA or by supplements, is really going to require some further observations.

Foggo, Edinburgh Can I ask Professor Hautvast what he thinks the role of N-3 fatty acids will be in the nineties?

Hautvast There are many studies going on. We are very concerned about taking too much, and are trying to define toxicological doses and also whether they have beneficial effects. I do not think it is the issue of the future, we feel after ten years' study that it is not so fascinating.

Rosenberg I might just add that while we were talking about antioxidant nutrients and so forth, one of the effects of feeding a high fish oil diet, at least to our elderly subjects, is to drive down vitamin E levels and also to have a detrimental effect on some immune functions, so that again we still have some more work to do to decide what doses of omega-3 and fish oils are going to be healthful as opposed to the other potential effects.

Questioner unknown Professor Hautvast, I realize that we have some agreement on agricultural policy for economic reasons and that nutritional quality is coming into our food policy, but where do you actually see nutrition coming into agricultural policy? Can you expand on that and give us some concrete examples?

Hautvast My feeling is that there is a growing interest in getting better agricultural policies. There is now discussion, for example, on the need to grow less sugar in Europe because less is eaten. Another example of agricultural policy with regard to cow rearing and butter is that we buy milk on the fat content and we may need to pay for milk by less butter fat and not by more butter fat. These things happen slowly, but possibly faster now than ten years ago, because the pressure for quality has never been so great. The agriculturists are willing to comply, despite being conservative by nature. They want to be sure of what to grow, but if they are convinced they will change for the better and produce food according to demand.

Davidson, Dundee Just a couple of questions for Professor Wharton on a practical level. You showed us a slide of some delightful foods which were very high in fibre, and yet you later went on to show another slide of a child who developed malnutrition due to a high fibre, low fat diet. Could you perhaps clarify for me whether we should be advising mums to introduce high fibre foods during the weaning process. Second, you were suggesting that milk should not be introduced to the diet of a child until they were over 1 year old, would this also include milk products such as yoghurt which are often used as weaning foods?

Wharton No, I do not think there should be any conscious attempt to introduce fibre specifically into the diet of a weaning child, because of energy density and considerations like that. Of course, some of the weaning foods that will be used contain fibre, but I do not think there should be an aim to include them particularly. As the first weaning food common throughout most of Europe is often rice you are going to have a certain amount of fibre. I am not absolutely sure what the effect of yoghurt is on the sort of things I have been talking about. Obviously yoghurt does not contain the iron needed, but there is no objection to that as such, as long as we do get some food which has available iron in it. I stress available iron and that is the point about iron-fortified formulae – it is very available iron. But we should not avoid yoghurt for any other reason so I have no objections to yoghurt (etc.) in weaning foods.

Cannon, London I notice that the introduction to this conference says 'Consumers perceive dietary advice as perplexing and confusing'. One could say that those who attend conferences on nutrition find the observations of nutritional scientists perplexing and confusing. The introduction to this conference also says 'There is a controversy as to what makes for a healthy diet' – as distinct from nutritional subfractions of the diet. But what healthy diet are we all talking about? Thirteen years ago Dr Passmore and others published 'Prescription for a better British diet' in the *British Medical Journal*. Is that it? We have not discussed it. Seven years ago the NACNE report was published and Professor Garrow commends this. Is this the better diet that we are looking for? It has not been discussed. Six years ago the COMA report on 'Diet and Cardiovascular Disease' was published. But there has been no discussion of it here. In the last two years in the United States we have had two colossal reports from the Surgeon General on 'Nutrition and Health' and from the National Academy of Sciences on 'Diet and Health'; both immense collaborations showing that the views of nutritional scientists all round the world on what constitutes a healthy diet are concordant. There is a consensus among scientists worldwide; but there has been no discussion here of the reports from the United States, although Dr McCarron has mentioned them. Two years ago the WHO published its report for Europe: 'Healthy Nutrition'. There has been no discussion on this here. This very week in Scotland we have the report on what constitutes a healthy diet for Scots. Here it is 'Prevention of Coronary Heart Disease in Scotland'. It set targets: fats, eventually 30% of total calories; saturated fats 10%, reduction in sugar and salt; an increase in bread, cereals, fruit and vegetables. Is that the prescription for a better diet in this country? Again it has not been discussed. I have to say that those of us who are concerned to institute a rational and progressive food and nutrition policy in this country are going to have to do without the advice of nutritional scientists. I beg the speakers this afternoon to address this issue.

McCarron I would be happy to answer this. I think good nutritional policy is based upon good science. As I think we heard yesterday, Dr Jackson said that as he joined the COMA committee he was impressed by the paucity of data. We have been trying to set a nutrition policy without having done the science to give us a proper data base to work from. Epidemiology only carries you so far. You must have properly controlled intervention trials. They are time consuming, they are costly, and the human subject is the most difficult laboratory animal to work with. We do not have the integration of good basic science with the nutritional sciences in terms of epidemiology and clinical trials. People are trying to do that. Dr Rosenberg runs a group at Tufts which is probably doing as good a job as any in the States. It is a challenge. People want to simplify things and maybe we are in an area that we cannot simplify as much as society wants. Dr Carlton, the editor of *Science*, at the end of 1989 wrote a very compelling editorial saying Western societies want a risk-free society, they want a simple message and unfortunately science cannot deliver that simple message because it is not simple. As for the Surgeon General's report, the National Academy of Science, the Diet and Health Committee and Food and Nutrition Board, their efforts are gallant. These reports identify important studies, but they fail to point out other equally important but sometimes controversial studies because those reports are put together by committees and often not by the people who actually did the research. That means that they do not have first-hand knowledge of the science that was executed. Looking at the NHANES data base, my conclusion is that people in America who eat a balanced diet, eat substantial quantities of food and do not drink alcohol appear to be the healthiest and leannest people in the country.

Appendix: Participants

Ms Ruzana Abdullah, Philip Henman Hall, Mylnes Court, Lawnmarket, Edinburgh EH1 2P

Mr K.G. Anderson, Brooke Bond Foods Ltd, Leon House, High Street, Croydon CR9 1JQ

Dr B.C.N. Ang, Department of Human Nutrition, London Hospital Medical College, London EC1V 2AD

Miss Ursula Arens, Nutritionist, Roche Products, Welwyn Garden City, Herts AL7 3AY

Ms Julia Armour, Forum Foods, 41–51 Brighton Road, Redhill, Surrey RH1 6YS

Ms Caroline Ashe, Account Manager, The Rowland Company, 67–69 Whitfield Street, London W1A 4PU

Ms Pauleen L. Auty, Senior Lecturer, Department of Applied Science, Leeds Polytechnic, Calverley Street, Leeds LS1 3HE

Dr I. Barclay, Lecturer, Food Technology Department, Auchincruive, Ayr KA6 5HW

Prof D.J.P. Barker, Director, Environmental Epidemiology Unit, Southampton General Hospital, Southampton SO9 4XY

Dr Margo E. Barker, Nutritionist, Dairy Council for Northern Ireland, 456 Antrim Road, Belfast BT15 8GB

Dr D.G. Beevers, Department of Medicine, Dudley Road Hospital, Birmingham B18 7QH

Dr Neville R. Belton, Senior Lecturer, Dept of Child Life and Health, 17 Hatton Place, Edinburgh EH9 1UW

Prof and Mrs Arnold E. Bender, 2 Willow Vale, Fetcham, Leatherhead, Surrey KT22 9TE

Dr Philip Beresford, Fisons PLC, 12 Derby Road, Loughborough, Leics LE11 0BB

Mr Ferry Biedermann, News Editor, 15 Belgrave Square, London SW1X 8PS

Dr Nino M. Binns, Manager, Chemical Products Registration, Pfizer Central Research, Sandwich, Kent CT13 9HN

Mrs M. Blades, Denehurst, 202 Newton Road, Rushden, Northants NN10 0SY

Dr C. Bolton-Smith, Cardiovascular Epidemiology Unit, Ninewells Hospital, Dundee DD1 9SG

Mrs Sarah Bond, Nutrition Department, Basset Crescent East, Southampton University, Southampton SO9 3TU

Mrs Shirley Bond, 15 Woodlands Drive, Yarm, Cleveland TS15 9NU

Ms B. Brodie, Gastrointestinal Laboratories, Western General Hospital, Crewe Road, Edinburgh EH4 2XU

Ms Helen Brown, Community Dietitian, City Health Clinic, Wellington Street, Peterborough PE1 3DU

Mrs N.M. Brunoro Costa, Flat 9, Creighton Court, Northcourt Avenue, Reading RG2 7RN

Mr W. Gordon Brydon, Principal Biochemist, Gastro Intestinal Laboratory, Western General Hospital, Crewe Road, Edinburgh EH4 2XU

Mrs Ann Burgess, Craiglea Cottage, Glenisla, Blairgowrie PH11 8PS

Dr Leslie Burgess, Craiglea Cottage, Glenisla, Blairgowrie PH11 8PS

Ms Carol Bushby, Regional Catering Adviser, Tayside Regional Council, Tayside House, Crichton Street, Dundee DD1 3RJ

Dr Mary Campbell Brown, 15 Trinity, Lynedoch Place, Glasgow G3 6AA

Miss Penelope Candry, Nutrition Health Promotion Officer, 58 Cecil Street, Lincoln LN1 3AT

Mr G.J. Cannon, 6 Aldridge Road Villas, London W11 1BP

Dr Eleanor Carlson, Nutritionist and Managing Director, Lifeline Nutritional Services Ltd, Pond House, 21 Cravenhill, London W2 3EN

Mrs M. Chinnery, Principal Lecturer, Food Science and Nutrition, Trinity and All Saints College, Brownberrie Lane, Horsforth, Leeds LS18 5HD

Mrs Brenda Clark, Chief Dietitian, Royal Hospital for Sick Children, Yorkhill, Glasgow G3 8SJ

Ms Valerie Clegg, Burson-Marsteller, 24–28 Bloomsbury Way, London WC1A 2PX

Mr S.B.L. Coles, Chelmshoe Farms, Great Madlestead, Halstead, Essex CO9 2RL

Ms Brenda Colledge, Senior Dietitian, Preston Hospital, North Shields, Tyne and Weir NE29 0LR

Dr Hazel Connors, United Biscuits UK, Group Research and Development Centre, Lane End Road, Sands, High Wycombe HP12 4JX

Miss H. Coubrough, Adviser in Dietetics, Dietetic Department, Royal Infirmary, Aberdeen AB9 22B

Ms Helen Crawley, ACLS, Polytechnic of North London, Holloway Road, London N7 8DB

Mrs R.E.A. Cruickshank, 68 North Gyle Terrace, Edinburgh EH12 8JY

Ms Alison Cullen, Nutrition and Dietetic Department, Royal Victoria Infirmary, Queen Victoria Road, Newcastle Upon Tyne NE1 4LP

Dr J.H. Cummings, Dunn Clinical Nutrition Centre, 100 Tennis Court Road, Cambridge CB2 1QL

Ms Mary Cursiter, Queen Margaret College, Clerwood Terrace, Edinburgh EH12 8TS

Ms Vivienne Davidson, Community Dietitian, Westgate Health Centre, Charleston Drive, Dundee

Mrs Jan Dawson, District Dietitian, The Royal Oldham Hospital, Rochdale Road, Oldham, Lancs OL1 2JH

Dr Anne de Looy, Principal Lecturer, Nutrition and Dietetics, Dept of Applied Science, Leeds Polytechnic, Leeds LS1 3HE

Dr Elizabeth Dodsworth, Managing Editor, Bureau of Nutrition, CAB International, Wallingford, Oxon OX10 8DE

Mr Keith Downton, Educational Catering Organiser, Lothian Regional Council, Department of Education, 40 Torphicen Street, Edinburgh EH3 8JJ

Dr Martin Eastwood, Consultant Physician, Gastrointestinal Unit, Western General Hospital, Edinburgh EH4 2XU

Dr Christine Edwards, Research Associate, Gastrointestinal Laboratories, Western General Hospital, Crewe Road, Edinburgh EH4 2XU

Miss J. Farquhar, Technician, Gastrointestinal Laboratories, Western General Hospital, Crewe Road, Edinburgh EH4 2XU

Mrs Karen Farquhar, Senior Dietitian, c/o Dietetic Department, Stracathro Hospital, Brechin, Angus DD9 7QA

Dr G.S. Fell, Institute of Biochemistry, Royal Infirmary, Glasgow G61 2JB

Dr C. Fenn, Lecturer, Robert Gordon Institute of Technology, Queens Road, Aberdeen AB9 2PG

Mr Simon Fevre, Senior Dietitian, Dietetic Department, Southern General Hospital, Glasgow G51 4TF

Mrs Margaret Foggo, Senior Dietitian, 14A Greenhill Terrace, Edinburgh EH10 4BS

Miss Lindsey Forrest, 39 Warwick Sreeet, Leicester LE3 5SE

Mrs Janice Fry, Chief Dietitian, Dept of Nutrition and Dietetics, Stirling Royal Infirmary, Stirling FK8 2AU

Ms Smita Ganatra, Nutritionist, Milk Marketing Board, Thames Ditton, Surrey KT7 0EL

Mr Alex Gardner, Head of School, Human Nutrition and Dietetics, The Queen's College, Glasgow G3 6LP

Prof J.S. Garrow, Department of Human Nutrition, Medical College of St Bartholomew's Hospital, Charterhouse Square, London EC1M 6BQ

Miss Susan J. Gatenby, Lecturer in Nutrition, Dept of Biochemistry, University of Surrey, Guildford GU2 5XH

Miss Samantha Gibbens, Researcher, Queen Margaret College, Edinburgh

Ms Anne Gibson, Community Health Services, Path House, 7 Nether Street, Kirkaldy, Fife KY1 2PG

Miss Paula A. Gilbert, Head Dietitian, Champneys Health Resort, Wigginton, Tring, Herts HP23 6HY

Ms Elizabeth Gilmour, Department of Surgery, The Medical School, Framlington Place, Newcastle Upon Tyne NE2 4LP

Ms E.V. Green, 142 Townehead Road, Fulham, London SW6 2RH

Ms Geraldine Grove, Nutrition Department, University of Southampton, Basset Crescent East, Southampton SO9 3TU

Mrs F. Hadley, Her Majesty's Inspector Of Schools, 55 Leigh Road, Walsall, West Midlands WS4 2DT

Ms Ruth Hailwood, Nutrition Department, University of Southampton, Basset Crescent East, Southampton SO9 3TU

Ms Penny Halford, Nutrition Department, University of Southampton, Basset Crescent East, Southampton SO9 3TU

Ms Jacqueline Hall, Nutritionist, CWS Technical Group, 28 Knowsley Street, Manchester M8 8JU

Ms Anne Halliday, Nutrition Scientist, British Nutrition Foundation, 15 Belgrave Square, London SW1X 8PG

Ms Linda Hanlon, GI Unit, Western General Hospital, Edinburgh EH4 2XU

Dr J.I. Harland, Head of Nutrition, British Sugar PLC, Peterborough PE2 9QU

Miss J. Harriman, St John's Hospital, Howden, Livingston, West Lothian

Prof J.G.A.J. Hautvast, Department of Human Nutrition, PO Box 8129, 6700 EV Wageningen, Netherlands

Dr K.W. Heaton, Reader in Medicine, Bristol Royal Infirmary, Bristol BS2 8HW

Ms Judith Hendry, Unit Dietitian, Inverurie Hospital, Inverurie AB5 9UL

Dr Steve Hewitt, Medical Services Manager, Fisons Consumer Health, 12 Derby Road, Loughborough, Leics LE11 0BB

Ms Kate Holborow, Director, Wearne Public Relations, 8 Southampton Place, London WC1A 2EA

Ms Carol A. Humphreys, 41 Lammas Court, St Leonard's Road, Windsor, Berks SL4 3ED

Ms Rosemary Hunt, District Food Policy Co-ordinator, Health Education Authority, Hamilton House, Mableton Place, London WC1H 9TX

Miss Juliet Hutton, Community Dietitian, 83 Rosemount Viaduct, Aberdeen AB1 1NS

Prof A.A. Jackson, School of Biochemical and Physiological Sciences, Medical and Biological Sciences Building, Bassett Crescent East, Southampton SO9 3TU

Prof and Mrs W.P.T. James, Director, The Rowett Research Institute, Greenburn Road, Bucksburn, Aberdeen AB2 9SB

Dr Eleri Jones, Lecturer, 62 Newborough Avenue, Llanishen, Cardiff, S. Glamorgan CF4 5DB

Ms Elaine Keegan, 63 Ramsbury Walk, Trowbridge, Wilts BA14 0UX

Ms Heather Kelman, Unit Dietitian, Inverurie Hospital, Inverurie AB5 9UL

Prof Robert E. Kendell, Dean, Faculty of Medicine, University of Edinburgh, Medical School, Teviot Place, Edinburgh

Mr Michael Kipps, Senior Lecturer, Dept of Management Studies, University of Surrey, Guildford, Surrey GU2 5XH

Ms Roslin Kirk, Principal Officer, Pre 5s Plus, Tayside Region Health Board, Dundee

Ms Alison Kirkby, Community Dietitian, Brierthorpe, 163 Durham Road, Stockton-on-Tees, Cleveland TS19 0EA

Dr D. Kritchevsky, Associate Director, The Wistar Institute of Anatomy and Biology, 36th and Spruce Streets, Philadelphia, Pennsylvania

Ms Jacquenne Landman-Bogues, Senior Lecturer, Department of Applied Science, Leeds Polytechnic, Calverley Street, Leeds LS1 3HE

Dr Michael E.J. Lean, Department of Human Nutrition, University of Glasgow, Glasgow Royal Infirmary, High Street, Glasgow

Ms Susan Lee, Information Scientist, Leatherhead Food RA, Randalls Road, Leatherhead, Surrey KT22 7RY

Mrs Jill Leslie, 1 Langford Mill, Mill Lane, Langford, Beds SG18 1QB

Ms Susan Love, Dietetic Department, East Birmingham Hospital, Boardsley Green East, Birmingham

Miss Jacqueline Lowdon, 42 Muircroft Terrace West, Burghmuir, Perth, PH1 1EB

Ms Catherine Lowe, Nutrition Programme Officer, Health Education Authority, Hamiliton House, Mabledon Place, London WC1H 9TQ

Ms Carole Lowis, Institute of Physiology, University of Glasgow, Glasgow G12 8QQ

Dr E.K. Lund, Institute of Food Research, Norwich Laboratory, Colney Lane, Norwich NR4 7UA

Dr D.A. McCarron, Nephrology and Hypertension, L463, Oregon Health Sciences, University, 3181 SW Sam Jackson Park Road, Portland, OR 97201

Ms Irene McClelland, Dept of Human Nutrition, Biological and Physical Sciences Building, Basset Crescent East, Southampton SO2 3TU

Ms Louise McCorkindale, Community Dietitian, Centre For Health Promotion, 39 Billing Road, Northampton NN1 5BA

Mrs Helen McDonald, Community Dietitian, Russell Institute, Causeyside Street, Paisley PA1 1UR

Prof. Ian Macdonald, Hillside, Fountain Drive, London SE19 1UP

Ms Elizabeth M. Macintosh, Home Economics Manager, Scottish Milk Marketing Board, Underwood Road, Paisley PA3 1TJ

Ms Sheila Macnaughton, Lecturer in Dietetics, The Queen's College, 1 Park Drive, Glasgow G3 6LP

Ms Jennifer Mair, Chief Dietitian, Dietetic Department, Southern General Hospital, Glasgow G51 4TF

Ms Y. Malcolm, Dietitian, Inverclyde Royal Hospital, Larkfield Road, Greenock PA16 0XN

Ms Karen Manson, Scientific Information Officer (Nutrition), CAB International, Wallingford, Oxon OX10 8WE

Mr D.J. Mela, Head of Sensory Studies, AFRC Institute of Food Research, Shinfield, Reading, Berks RG2 9AT

Mrs Alice Michie, Senior Dietitian, Dept of Dietetics, Western General Hospital, Edinburgh EH4 2XU

Miss Cathy Mooney, Senior Lecturer, Nutrition and Dietetics Department, Polytechnic of North London, Holloway Road, London N7 8DB

Prof H. Gemmell Morgan, Firwood House, 8 Eaglesham Road, Newton Mearns, Glasgow G77 5BG

Dr Edmund J. Moynahan, The Malting House, Arkesden, Saffron Walden, Essex, CB11 4HB

Miss Jane Murphy, Nutrition Department, School of Biological and Physical Sciences, Basset Crescent East, Southampton SO9 3TU

Dr Wilson M. Nicol, Eurobest Associates, Principal, 83A Kidmore Road, Caversham, Reading RG4 7NQ

Dr John O'Brien, Nutrition Editor, Elsevier

Miss Helen Orr, Dykehead Farm, Chapelton, Strathaven, Lanarkshire, ML10 6SL

Mr Robin M. Orr, Lecturer, Seale Hayne Faculty of Food, Polytechnic South West, Newton Abbot, Devon TQ12 6NQ

Prof. Doreen A. Parry, Head of Department, Dietetics and Nutrition, Queen Margaret College, Edinburgh EH12 8TS

Dr R. Passmore, 54 Newbattle Terrace, Edinburgh EH10

Ms Anne Payne, Dept of Child Life and Health, 17 Hatton Place, Edinburgh EH9 1UW

Dr Jack Pearce, Food and Agricultural Chemistry Dept, Queen's University, Newforge Lane, Belfast BT9 5PX

Ms Victoria Pennington, Farley Health Products, Nutrition Department, Consumer Products Development, Nottingham NG2 3AA

Mr John Perry, Senior Education Officer, Lothian Regional Council, Education Department, 40 Torphicen Street, Edinburgh EH3 8JJ

Mrs M. Pinder, 2 Abbey Street, Cerne Abbas, Dorchester Dorset

Dr J. Powell-Tuck, Department of Human Nutrition, London Hospital Medical College, London EC1V 2AD

Ms Jane A. Pryer, 77 Slinn Street, Sheffield S10 1NW

Dr Fatemeh Rabiee, Nutritionist, Health Education Department, St Patricks, Highgate Street, Birmingham B12 0YA

Ms Catherine Ravenscroft, 28 Porson Road, Cambridge CB2 2EU

Dr Francis B. Reed, Biochemistry Division, UMDS, Guy's Hospital, London, SE1 9RT

Mrs Ruth C. Richards, Consultant Nutritionist, 5 Ethendun, Bratton, Westbury, Wilts BA13 4RX

Dr John Richmond, President, Royal College of Physicians, 9 Queen Street, Edinburgh EH2 1JQ

Miss Aileen Robertson, Chief Dietitian, Dept Nutrition and Dietetics, Raigmore Hospital, Inverness

Dr Paul J. Roderick, 18 Halfacre Road, Hanwell, London W7 3JJ

Ms Helen Rose, Room 205, Nobel House, Ministry of Agriculture, Fisheries and Food, Smith Square, London SW1P 3HX

Dr I.H. Rosenberg, Director, Human Nutrition Research Center, Aging at Tufts University, 711 Washington Street, Boston, Massachusetts 02111 USA

Dr David Rumsey, Centre For Human Nutrition, University of Sheffield, Sheffield S10 2TA

Ms Margaret Sanderson, Dept of Applied Chemistry and Life Science, Polytechnic of North London, Holloway Road, London N17 8DB

Ms Claire Schofield, Nutritionist, 29 Perowne Street, Cambridge CB1 2AY

Mrs Claire Seaman, Dept of Dietetics and Nutrition, Queen Margaret College, Edinburgh EH12 8TS

Dr A. Shenkin, Institute of Biochemistry, Royal Infirmary, Glasgow G4 0SF

Prof J Shepherd, Department of Pathological Biochemistry, Royal Infirmary, Glasgow G4 0SF

Miss Colette Shortt, Research Nutritionist, Rowett Research Institute, Aberdeen

Ms Catherine J. Smith, Snr Research and Development Officer, SHEG Woodburn House, Canaan Lane, Edinburgh EH10 4SG

Dr Alison M. Stephen, Assistant Professor, Division of Nutrition and Dietetics, College of Pharmacy, University of Saskatchewan, Saskatoon S7N OWO

Dr and Mrs E.A. Stevens, Great Cooting, Adisham, Nr Canterbury, Kent CT3 3JR

Ms Lynn Stockley, Nutritionist, Health Education Authority, Hamilton House, Mabledon Place, London WC1 9TX

Mrs Maureen Strong, Faculty of Catering, Cornwall College, Pool, Cornwall TR15 3RD

Mr K. Sugden, Group Head, Development, Reckitt and Colman Pharmaceuticals, Dansom Lane, Kingston-upon-Hull, HU8 7DS

Mr J.C.K. Wells, Paediatric Nutrition Manager, Cow and Gate Ltd, Manvers Street, Trowbridge, Wiltshire BA14 8YX

Mr R.W. Wenlock, Nutritionist, Department of Health, Alexander Fleming House, Elephant and Castle, London SE1 6BY

Prof. B.A. Wharton, Dept of Human Nutrition, University of Glasgow, Yorkhill Hospital, Glasgow G3 8SJ

Dr Christine M. Williams, Lecturer in Nutrition, Dept of Biochemistry, University of Surrey, Guildford GU2 5XH

Ms Pamela J. Wilson, Dietitian, City General Hospital, Newcastle Road, Stoke-on-Trent ST4 6QG

Mr Jack Winkler, 28 St Paul Street, London N1 7AB

Dr A. Wise, Robert Gordon's Institute of Technology, Queen's Road, Aberdeen AB9 2PG

Dr M. Wiseman, Head of Nutrition Unit, Room C307, Department of Health, Alexander Fleming House, London SE1 6BY

Dr John Wright, Department of Biochemistry, University of Surrey, Guildford GU2 5XH

Dr K.C. Yates, Manager, Scientific and Consumer Affairs, Kellogg Company of Great Britain, Park Road, Stretford, Manchester M32 8RA

Ms Eileen Young, Community Dietitian, Russell Institute, Causeyside Street, Paisley PA1 1UR

Index